How to Solve Algebra Word Problems

How to Solve Algebra Word Problems

WILLIAM A. NARDI

Prentice Hall
New York • London • Toronto • Sydney • Tokyo • Singapore

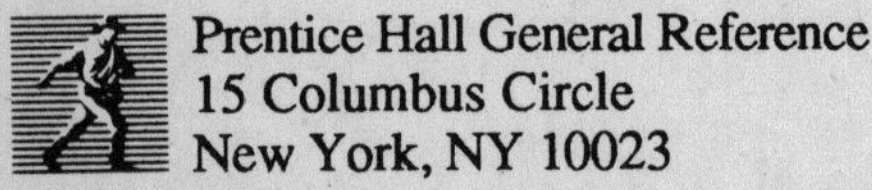
Prentice Hall General Reference
15 Columbus Circle
New York, NY 10023

An Arco Book

Library of Congress Cataloging-in-Publication Data

Nardi, William A.
How to solve algebra word problems / William A. Nardi
p. cm.
ISBN 0-13-425216-0
1. Algebra—Problems, exercises, etc. 2. Word problems
(Mathematics). I. Title.
QA157.N37 1991 91-13800
512—dc20 CIP

Manufactured in the United States of America

4 5 6 7 8 9 10

Contents

Foreword 1

1 Translating English Terms Into Algebra Symbols 3

2 Translating English Statements Into Algebra Equations 8

3 Solving Equations 10

4 Distance, Rate, and Time 14

5 Coin Problems 23

6 Mixture Problems 30

7 Investment Problems 45

8 Work Problems 52

9 Attendance Problems 64

10 Lever Problems 67

11 Age Problems 70

12 Ratio Problems 75

13 Proportion 78

14 Inequalities 83

15 Solving Problems with Two Variables 92

16 Word Problems Review Test 117

17 Problems Involving Quadratic Equations 147

18 Trigonometric Word Problems 160

19 Probabilities, Permutations, and Combinations 168

Table of Squares and Square Roots 181

Table of Values of Trigonometric Functions 182

ACKNOWLEDGMENTS

I wish to thank my daughter, Nadine Nardi Davidson, for her editorial assistance and faith in this project; Vernon Manley for his helpful suggestions; and my wife, Anne, for her patience and encouragement.

William A. Nardi, Sr.

Foreword

Word problems—they continue to be the downfall of most algebra students. Yet, every student, regardless of his or her chosen field of study, will inevitably face them over and over again on college and graduate school entrance exams and even on job placement exams. Students seeking careers in mathematics, engineering, science, and related fields will need the techniques not only to solve these problems, but also to create them.

Even if you have experienced little difficulty solving problems when an equation is given, you may find yourself continually frustrated or slowed down when these problems are presented in words only. Why? Because word problems add an extra challenge—that of first setting up the equation, of translating the word problem from English into an algebraic equation before it can be solved.

This book is a road map that will help you make your way through the word-problem maze by teaching you to recognize and isolate the key words in typical problems and then to translate them into algebraic equations.

However, as algebra is the extension of arithmetic, it is assumed that all the fundamentals, including whole numbers, fractions, decimals, and percents, have already been mastered. The student should also have a knowledge of solving equations with one and two unknowns and with positive and negative integers.

1 Translating English Terms Into Algebra Symbols

The most important and difficult problems for the algebra student to solve are word problems. These problems often use different words to mean the same thing. To make it easier to solve these problems, some frequently used words and their mathematical translations are listed below.

After you have studied this list, turn to page 5. The same list appears but without the translations. Write in the translations in the blank spaces. Do not go on to page 6 until you can identify all the terms.

Term	Symbol of Translation
add	+
sum	+
more	+
increased	+
greater	+
excess	+
less	−
difference	−
subtract	−
diminished	−
reduced	−
remainder	−
product	×
divide	÷
find quotient	÷
quantity	( )
equal	=
result	=

Term	Symbol of Translation
less than	$<$
greater than	$>$
greater than or equal to	$\geq$
less than or equal to	$\leq$

Examples

Words	Symbol of Translation
1. Nine less than twelve.	$12-9$
2. Four less than three times y.	$3y-4$
3. Six less two.	$6-2$
4. Four times the quantity six plus two.	$4(6+2)$
5. One-half a number times fifteen.	$\frac{1}{2}n(15)$
6. One-third a number less two.	$\frac{1}{3}n-2$
7. Ten times the sum of twice a number and six.	$10(2n+6)$
8. Eighteen times the difference of a number and ten.	$18(n-10)$
9. Eight times the quantity y plus two divided by four.	$\frac{8(y+2)}{4}$

TEST YOURSELF

Term	Symbol of Translation
add	________
sum	________
more	________
increased	________
greater	________
excess	________
less	________
difference	________
subtract	________
diminished	________
reduced	________
remainder	________
product	________
divide	________
find quotient	________
quantity	________
equal	________
result	________
less than	________
greater than	________
greater than or equal to	________
less than or equal to	________

TRANSLATING ENGLISH TERMS INTO ALGEBRA SYMBOLS

ENGLISH (use n to represent the number)	ALGEBRA
1. A number increased by six.	**1.** $n+6$
2. Twice a number decreased by six.	**2.** $2n-6$
3. Four less than five times a number.	**3.** $5n-4$
4. The sum of three times a number and twelve.	**4.** $3n+12$
5. One-half a number is fifty.	**5.** $\frac{1}{2}n=50$
6. If from twice a number you subtract four, the difference is twenty.	**6.** $2n-4=20$
7. Twenty-five is nine more than four times a number.	**7.** $25=9+4n$
8. Sixteen subtracted from five times a number equals the number plus four.	**8.** $5n-16=n+4$
9. Twenty-five is the same as ten added to twice a number.	**9.** $25=10+2n$
10. Three x is five less than twice x.	**10.** $3x=2x-5$
11. The sum of two y and three is the same as the difference of three y and one.	**11.** $2y+3=3y-1$
12. The difference of five and five y is the same as eight and two y.	**12.** $5-5y=8+2y$
13. Twice the quantity of two y and six.	**13.** $2(2y+6)$
14. Twice the quantity of seven plus x is the same as the difference of x and seven.	**14.** $2(7+x)=x-7$
15. The sum of two y and the quantity of three plus y plus twice the quantity two y minus two equals fifteen.	**15.** $2y+(3+y)+2(2y-2)=15$

English–Algebra Translation Test

Try translating the following from English to algebra (let n stand for the number). Then check your answers with those below. If you missed more than 3, restudy pages 3–6. and then try again.

ENGLISH (use n to represent the number)	ALGEBRA
1. A number added to five.	**1.** ________
2. The difference between six n and seven.	**2.** ________
3. One-half a number is thirty.	**3.** ________
4. Twice a number less twelve.	**4.** ________
5. Sixty-four diminished by n equals one.	**5.** ________
6. A number is equal to fifty less the number.	**6.** ________
7. Sixteen subtracted from five times a number equals the number plus four.	**7.** ________
8. Twice a number decreased by eight is zero.	**8.** ________
9. A number is equal to fifty less nine times the number.	**9.** ________
10. The product of twice a number and sixteen is sixty-four.	**10.** ________
11. If the sum of six and a number is divided by two, the quotient is twelve.	**11.** ________
12. If the product of five and a number is divided by six the result is ten.	**12.** ________
13. Twenty-five is five less than six n diminished by two.	**13.** ________
14. The quotient of sixteen n divided by one hundred equals sixteen.	**14.** ________
15. Twice the quantity of $n+$two divided by six is fifty.	**15.** ________

Answers to English–Algebra Translation Test

1. $n+5$

2. $6n-7$

3. $\frac{1}{2}n=30$

4. $2n-12$

5. $64-n=1$

6. $n=50-n$

7. $5n-16=n+4$

8. $2n-8=0$

9. $n=50-9n$

10. $2(n+16)=64$

11. $\frac{(6+n)}{2}=12$

12. $\frac{5n}{6}=10$

13. $25=(6n-5)-2$

14. $\frac{16n}{100}=16$

15. $\frac{2(n+2)}{6}=50$

2 Translating English Statements Into Algebra Equations

Many word problems can be simplified by using the five hints listed below:

1. Read the problem at least twice.

2. Draw a diagram if possible. If it is a geometric problem, this is a *must*.

3. Underline the variable (unknown). When a problem concerns two things or two people, whatever is mentioned last in the sentence is the unknown.

4. Write an equation relating to the numbers in the problem. Let x represent the unknown or variable.

Example 1

John's age is twice Mary's age.

Let x = Mary's age (Mary's age was mentioned last.)
$2x$ = John's age

Example 2

(Always draw the linear figure.)
The length of a rectangle is 6 more than twice the width (mentioned last, so equals x).

Let x = width
$2x + 6$ = length

x

2x + 6

Example 3

There are 5 times as many dimes as quarters.

Let x = quarters
$5x$ = dimes

Example 4

The rate of a plane is 10 times that of a freight train.

Let $x =$ rate of freight train
$10x =$ rate of plane

Example 5

Mr. Johnson invested 500 dollars more than twice Mr. Rodriguez's investment.

Let $x =$ Mr. Rodriquez's investment
$500 + 2x =$ Mr. Johnson's investment

5. When you have 3 or more types of coins or other items in a problem, use x to represent whatever is not mentioned.

Example 1

Wilfred has 46 coins amounting to \$6.50. The number of dimes is twice the number of nickels. The number of quarters is 6 more than the number of nickels. How many of each coin does he have?

Let $x =$ number of nickels
$2x =$ number of dimes
$(x+6) =$ number of quarters

Example 2

Harold has twice as much money as Nadine. Nadine has 3 times as much money as Sandra. If together they have \$357.00, how much money does each of them have? (Notice that in the above problem, Sandra's money is not compared as stated with either Harold's or Nadine's. We therefore assign x to her amount.)

Let $x =$ amount of money Sandra has
$3x =$ amount of money Nadine has
$6x =$ amount of money Harold has

Equation: $x + 3x + 6x = 357$
$10x = 357$
$x = \$35.70$ (Sandra)
$3x = \$107.10$ (Nadine)
$6x = \$214.20$ (Harold)

3 Solving Equations

Definition: When something equals something else, you have an equation.

Examples

a. $x=6$ **c.** $5n+2n-16=2n+14$

b. $x+5=6$ **d.** $6(n+6)=-2(3n-6)+n$

Some reminders:

Every equation has two members—a right and a left separated by an equal sign. When bringing (transposing) a term from a right member to a left member or from a left member to a right member, you must change the sign of the term from + (positive) to − (negative) or from − (negative) to + (positive).

Likewise, a term that is multiplied on one side is divided when moved or transposed to the opposite side, and a term that is divided on one side is multiplied on the opposite side.

To solve the equation, you first transpose like terms to the same side of the equation.

Example 1

$3x+6=12$	1. Transpose like terms to same side of equation.
$3x=12-6$	(When +6 is transposed, it becomes −6.)
$3x=6$	2. Collect like terms ($12-6=6$).
$x=6\div 3$	3. Transpose like terms.
	(When the 3 is transposed it is divided into the 6.)
$x=2$	Solution.

Example 2

$8x+16=5x+1$	1. Transpose like terms to same side of equation.
$8x-5x=-16+1$	(Remember to change signs when crossing the equal sign.)
$3x=-15$	2. Collect like terms.
$x=-15\div 3$	3. Transpose the 3 by dividing it into −15.
$x=-5$	Solution.

Example 3

$5(t-14)+3(t+6)=-2(t+5)-2$	
$5t-70+3t+18=-2t-10-2$	1. Remove parentheses.
$8t-52=-2t-12$	2. Collect like terms.

$8t+2t=+52-12$	3. Transpose like terms.
$10t=+40$	4. Collect like terms.
$t=40\div 10$	5. Transpose again.
$t=4$	Solution.

Practice Problems: Solving Equations

1. The sum of two numbers is 231. The larger is twice the smaller. What are the numbers?

2. The length of a rectangle is three times the width. If the perimeter is 80 feet, what are the dimensions?
(Reminder: The perimeter is the sum of all sides added together.)
Remember to draw the figure.

3. The length of a rectangle is 6 less than twice the width. If the perimeter is 60 inches, what are the dimensions? (Draw a figure.)

4. A man earns $1,000 more than 3 times what his son earns. If the total amount of what they both earn is $29,000, how much money did the father and son each earn?

5. A piece of wire 60 inches long is cut into 2 pieces so that the larger piece is 10 inches longer than the shorter piece. How long is each piece of wire?

6. Two tuna boats start from the same port at the same time, but they head in opposite directions. The faster boat travels 10 knots per hour faster than the slower boat. At the end of 8 hours, they were 272 nautical miles apart. How many nautical miles had each boat traveled by the end of the 8-hour period?

7.

x cm
x cm
2x cm
2x cm

The length of the larger square is twice as large as the length of the smaller square. If the sum of their perimeters is 648 centimeters, what is the length of each square?

8. In an algebra class there are 10 less than twice as many boys as girls. If the total number of the students is 38, how many boys and girls are there in the class?

9. The cost of a bottle and a cork is $1.10. If the bottle costs $1.00 more than the cork, what is the cost of each?

10. A man's will left $75,000 to his wife, son, and daughter in such a way that his son's share was 3 times the daughter's share, while the wife received as much as the son and daughter together. How much did each receive? Hint: Let $x=$ daughter's share.

Check your answers with the solutions that follow.

Answers to Practice Problems: Solving Equations

1. Let $x =$ smaller number
$2x =$ larger number

$$\begin{aligned} 2x + x &= 231 \\ 3x &= 231 \\ x &= 77 \text{ smaller number} \\ 2x &= 154 \text{ larger number} \end{aligned}$$

2. Let $x =$ width in feet
$3x =$ length in feet

x ft
3x ft

$$\begin{aligned} 2x + 6x &= 80 \text{ ft} \\ 8x &= 80 \\ x &= 10 \text{ ft width} \\ 3x &= 30 \text{ ft length} \end{aligned}$$

3. Let $x =$ width in inches
$(2x - 6) =$ length in inches

x in
2x – 6 in

$$\begin{aligned} 2x + 2(2x - 6) &= 60 \\ 2x + 4x - 12 &= 60 \\ 6x &= 72 \\ x &= 12 \text{ in width} \\ (2x - 6) &= 18 \text{ in length} \end{aligned}$$

4. Let $x =$ money son earns
$3x + 1{,}000 =$ money father earns

$$\begin{aligned} x + 3x + 1{,}000 &= 29{,}000 \\ 4x &= 29{,}000 - 1{,}000 \\ 4x &= 28{,}000 \\ x &= \$7{,}000 \text{ money son earns} \\ 3x + 1{,}000 &= \$22{,}000 \text{ money father earns} \end{aligned}$$

5. Let $x =$ length of shorter piece
$(x + 10) =$ length of larger piece

$$\begin{aligned} x + x + 10 &= 60 \text{ in} \\ 2x + 10 &= 60 \text{ in} \\ 2x &= 60 - 10 \\ 2x &= 50 \\ x &= 25 \text{ in for shorter piece} \\ x + 10 &= 35 \text{ in for larger piece} \end{aligned}$$

6. Let $x =$ knots per hour for slower boat
$(x+10) =$ knots per hour for faster boat
$8x =$ nautical miles traveled in 8 hrs by slower boat
$8(x+10) =$ nautical miles traveled in 8 hrs by faster boat

$$8x+8(x+10)=272$$
$$8x+8x+80=272$$
$$16x+80=272$$
$$16x=192$$
$x=12$ knots per hour for slower boat
$x+10=22$ knots per hour for faster boat
$8x=96$ nautical miles traveled by slower boat
$8x+80=176$ nautical miles traveled by faster boat

7. Let $x =$ length of side in cm for smaller square
$4x =$ length of perimeter in cm of smaller square
Let $2x =$ length of side in cm for larger square
$8x =$ length of perimeter in cm of larger square

$$4x+8x=648 \text{ cm}$$
$$12x=648 \text{ cm}$$
$x=54$ cm, length of side for smaller square
$2x=108$ cm, length of side for larger square
$4x=216$ cm, length of perimeter of smaller square
$8x=432$ cm, length of perimeter of larger square

8. Let $x =$ number of girls
$2x-10 =$ number of boys

$$x+2x-10=38$$
$$3x=38+10$$
$$3x=48$$
$x=16$ girls
$2x-10=22$ boys

9. Let $x =$ cost of cork
$x+\$1.00 =$ cost of bottle

$$2x+\$1.00=\$1.10$$
$$2x=\$1.10-1.00$$
$$2x=\$\ .10$$
$x=\$\ .05$ cost of cork
$x+\$1.00=\$1.05$ cost of bottle

10. Let $x =$ money received by daughter
$3x =$ money received by son
$4x =$ money received by wife

$$x+3x+4x=\$75{,}000$$
$$8x=\$75{,}000$$
$x=\$\ 9{,}375$ received by daughter
$3x=\$28{,}125$ received by son
$4x=\$37{,}500$ received by wife

4 Distance, Rate, and Time

A very important formula for solving these problems: $D = R \times T$

$D =$ Distance
$R =$ Rate
$T =$ Time

Learn it.
Also, learn these important formulas: $R = \frac{D}{T}$ $T = \frac{D}{R}$

1. If a car travels at an average rate of 50 miles per hour, then in two hours it will have traveled __?__ miles.
2. If a plane traveled 2,184 miles in 3.5 hours, then its rate of speed is equal to __?__ miles per hour.
3. If a passenger train travels at an average rate of 55 miles per hour, then it will take __?__ hours to travel 275 miles.

When solving motion problems, always draw a diagram and then use the model illustrated in the example that follows.

Example

Two cars start from the same point and travel in opposite directions. The rate of the slower car is 15 miles per hour less than the rate of the faster car. After 8 hours they are 840 miles apart. Find the rate of speed of each car.

Diagram:

Let $x =$ rate of faster car
$x - 15 =$ rate of slower car

Model:

Formula	*Distance*	=	*Rate*	×	*Hours*
Faster car	$8x$		x		8
Slower car	$8(x-15)$		$x-15$		8

The distance of the faster car plus the distance of the slower car = Total Distance.

Equation:

$$8x + 8(x - 15) = 840$$
$$8x + 8x - 120 = 840$$
$$16x = 840 + 120$$
$$x = 60 \text{ rate of faster car}$$
$$x - 15 = 45 \text{ rate of slower car}$$

STUDY PROBLEM 1

Two trains leave the same train station at the same time, but in opposite directions. The faster train travels at an average rate of 75 miles per hour and the slower train travels at an average rate of 68 miles per hour. In how many hours will they be 715 miles apart?

Diagram:

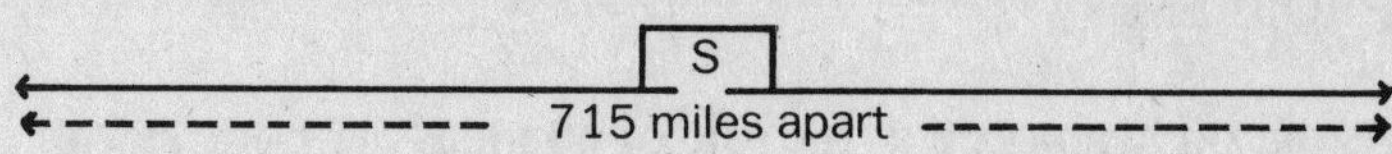

Model:

Formula	*Distance*	=	*Rate*	×	*Hours*
Faster train	$75x$		75		x
Slower train	$68x$		68		x

Distance of faster train $= 75x$
Distance of slower train $= 68x$
Total distance $= 143x$

Equation:

$$143x = 715$$
$$x = 5 \text{ hours}$$

STUDY PROBLEM 2

Mr. Granger, a cyclist, rode from his home to his office at the average speed of 18 miles per hour. On his return home from his office, using the same route, he averaged 12 miles per hour. If the total round trip took 5 hours, what was the distance from his home to his office?

Diagram:

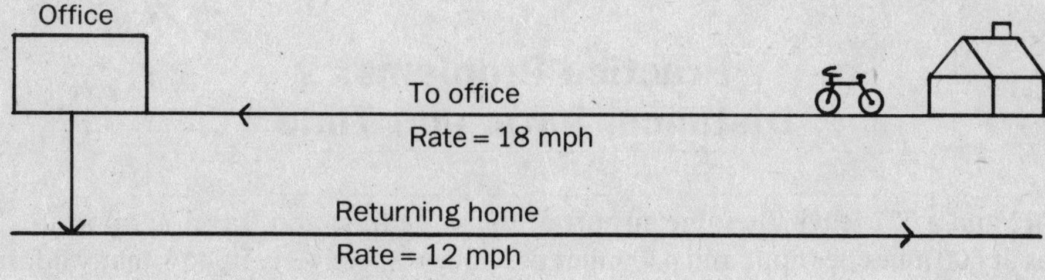

Model:

Formula	*Distance*	=	*Rate*	×	*Hours*
Home to office	$18x$		18		x
Office to home	$12(5-x)$		12		$5-x$

Equation:

$$18x = 12(5-x)$$
$$18x = 60 - 12x$$
$$18x + 12x = 60$$
$$30x = 60$$
$$x = 2 \text{ hours}$$
$$18x = 36 \text{ miles}$$

STUDY PROBLEM 3

Steve starts out on his moped at the rate of 40 miles per hour. Two hours later, Bill drives his car along the same route at 55 miles per hour. In how much time will Bill overtake Steve?

Diagram:

Model:

Formula	*Distance*	=	*Rate*	×	*Hours*
Moped	$40(x+2)$		40		$x+2$
Car	$55x$		55		x

Equation:

$$55x = 40(x+2)$$
$$55x = 40x + 80$$
$$55x - 40x = 80$$
$$15x = 80$$
$$x = 5\tfrac{1}{3} \text{ hours (5 hr and 20 min)}$$

Practice Problems: Distance, Rate, and Time

1. A 747 and a 727 leave the same airport at the same time and travel in opposite directions at 600 miles per hour and 450 miles per hour respectively. In how many hours will they be 3,150 miles apart?

Formula	*Distance*	=	*Rate*	×	*Hours*
?	?		?		?
?	?		?		?

Write equation. Answer is __?__ hours.

2. A helicopter traveling at an average rate of speed of 150 miles per hour left from a downtown terminal at noon after a train which had departed from the same terminal at 9 A.M. If the helicopter overtook the train at 2:00 P.M., find the average speed of the train.

Draw diagram.
Draw model.
Write equation.
Answer is __?__ hours.

3. Two cars 720 kilometers apart travel toward each other. One car travels 70 kilometers per hour while the other car travels at the rate of 50 kilometers per hour. In how many hours will they meet?

Hint: Let $x =$ hours it will take before they meet.
Distance covered by first car = 70 (the rate) × (time in hours).
Distance covered by second car = 50 (the rate) × (time in hours).
Equation: ______?______.
Answer is __?__ hours.

4. Two boys on bicycles start from the same place at the same time. One cyclist travels 15 miles per hour, the other travels 12 miles per hour. If they travel in the same direction, how many hours will it take before they are 9 miles apart?

Hint: Let $x =$ time it takes to be nine miles apart.
$15x =$ distance of first cyclist.
$12x =$ distance of second cyclist.
Equation: ______?______.
Answer is __?__ hours.

5. A plane leaves St. Louis airport at 10:00 A.M. flying due north. Another plane from the same airport flies due south at 12:00 P.M. At 2:00 P.M. they are 1,800 miles apart. Find the rate of speed of each plane if the rate of the plane flying north is twice that of the plane flying south.

Let $x =$ rate of plane flying south.
_?_$x =$ rate of plane flying north.
_?_$x =$ total rate of both planes.
Equation: ______?______.
Answer is __?__ for the plane flying south.
Answer is __?__ for the plane flying north.

6. A freight train takes 18 hours to travel the same distance that an express train travels in 15 hours. The rate of the express train is 15 miles per hour faster than the freight train. Find the rate at which each train travels.

Draw diagram.
Draw model.
Write equation.

Answer is ___?___ mph for the express train.
Answer is ___?___ mph for the freight train.

7. At 10:00 A.M. two classmates start out on their bikes to meet each other from towns located 68 miles apart. At 1:00 P.M. they meet. If one boy traveled 3 miles per hour faster than the other, what was the rate of speed of each boy?
   Draw diagram.
   Draw model.
   Write equation.
   Answer is ___?___ mph for the faster bike.
   Answer is ___?___ mph for the slower bike.

8. At 12:00 noon two Mississippi steamboats are 126 miles apart. At 6:00 P.M. they pass each other going in opposite directions. If one steamboat travels 9 miles per hour faster than the other, find the speed of both boats.
   Draw diagram.
   Draw model.
   Write equation.
   Answer is ___?___ mph for the faster boat.
   Answer is ___?___ mph for the slower boat.

9. A jet bomber traveling 500 miles per hour arrived over its target at the same time as its fighter jet escort which left the same base ½ hour after the bomber took off. How many hours did it take to reach the target if the fighter jet traveled 600 miles per hour?
   Draw diagram.
   Draw model.
   Write equation.
   Answer is ___?___ hours.

10. A private jet had been flying for 2 hours when it encountered strong head winds which reduced its speed by 40 miles per hour. If it took the plane 5 hours to travel 1,380 miles, find its speed before flying into the head winds.
    Hint: Let x = speed of jet during first 2 hours of flight.
    Draw diagram.
    Draw model.
    Write equation.
    Answer is ___?___ mph.

Check your answers with the solutions that follow. If you missed more than 3, review this chapter again before going on to Chapter 5.

Answers to Practice Problems: Distance, Rate, and Time

1.

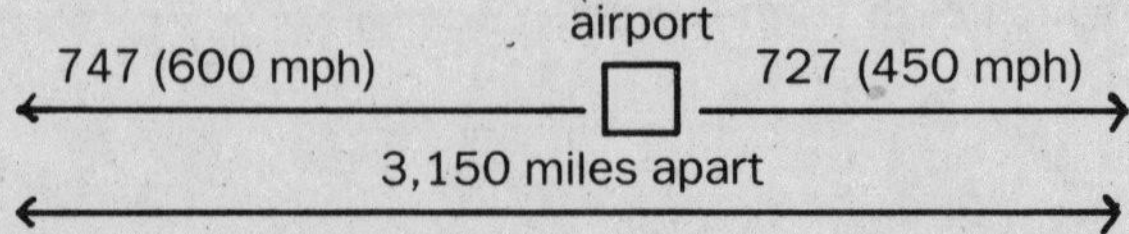

Formula	*Distance*	=	*Rate*	×	*Time*
747	$600x$		600		x
727	$450x$		450		x

Equation:

$$600x + 450x = 3{,}150 \text{ Total Distance}$$
$$1{,}050x = 3{,}150$$
$$x = 3 \text{ hr to be } 3{,}150 \text{ miles apart}$$

2.

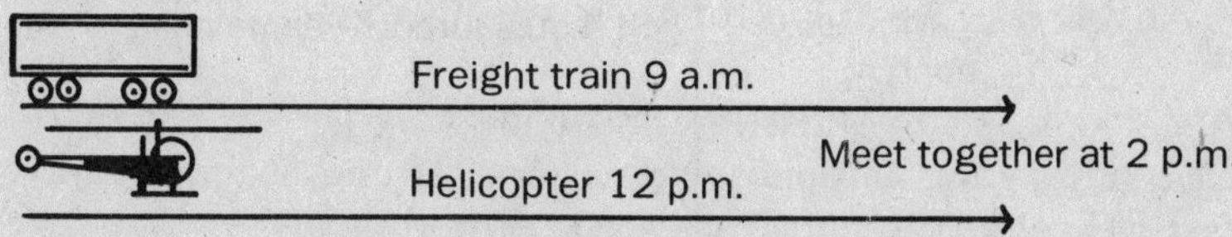

Formula	*Distance*	=	*Rate*	×	*Time*
Freight train	$5x$		x		5
Helicopter	300		150		2

Equation:

$$5x = 300$$
$$x = 60 \text{ mph} - \text{the average speed of train}$$

3.

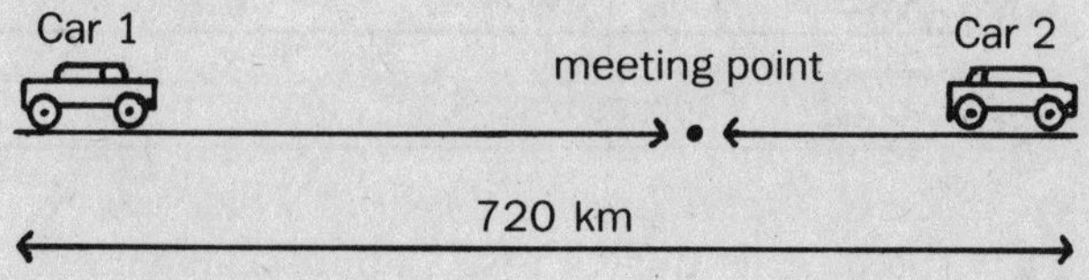

Formula	*Distance*	=	*Rate*	×	*Time*
Car 1	$70x$		70		x
Car 2	$50x$		50		x

Equation:

$$70x + 50x = 720 \text{ km}$$
$$120x = 720 \text{ km}$$
$$x = 6 \text{ hr before they meet}$$

4.

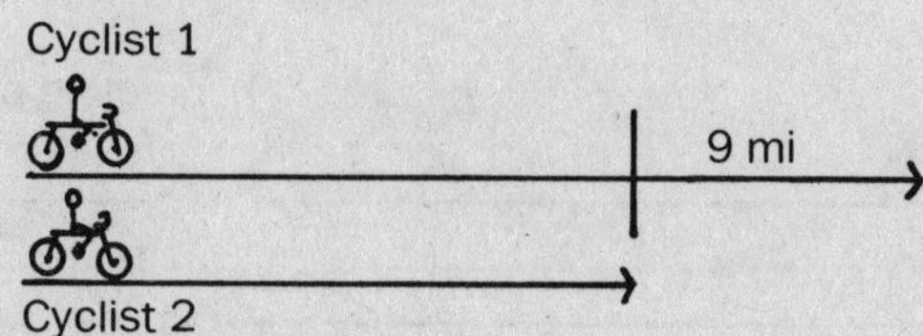

Formula	*Distance*	=	*Rate*	×	*Time*
1st boy	$15x$		15		x
2nd boy	$12x$		12		x

Equation:

$$15x - 12x = 9$$
$$3x = 9$$
$$x = 3 \text{ hr}$$

5.

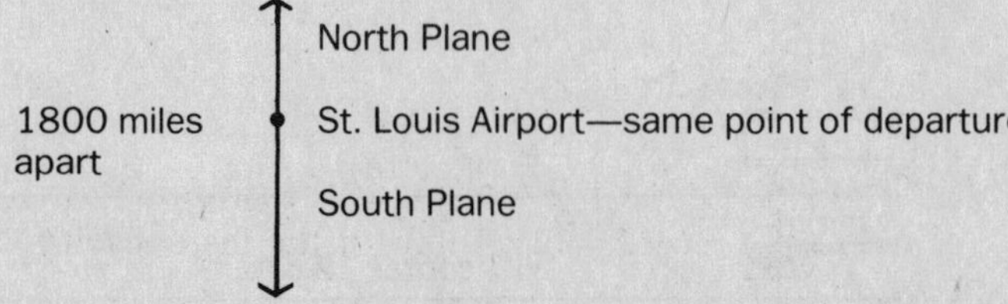

Formula	*Distance*	=	*Rate*	×	*Time*
North plane	$8x$		$2x$		4 hr
South plane	$2x$		x		2 hr

Equation:

$$8x + 2x = 1{,}800$$
$$10x = 1{,}800$$
$$x = 180 \text{ mph speed of plane flying south}$$
$$2x = 360 \text{ mph speed of plane flying north}$$

6.

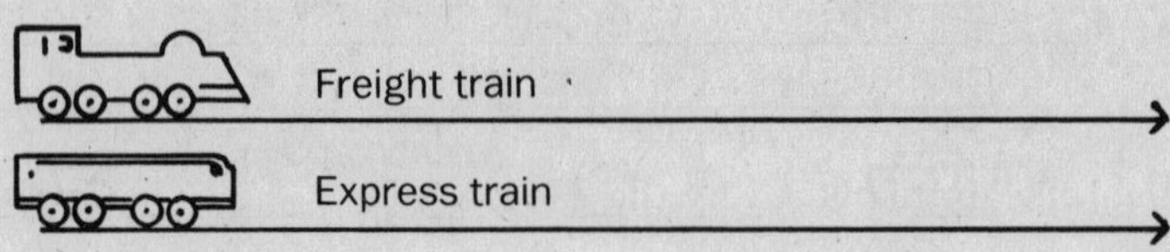

Formula	*Distance*	=	*Rate*	×	*Time*
Freight train	$18x$		x		18 hr
Express train	$15(x+15)$		$x+15$		15 hr

Equation:

$$18x = 15(x+15)$$
$$18x = 15x + 225$$
$$18x - 15x = 225$$
$$3x = 225$$
$$x = 75 \text{ mph for the freight train}$$
$$x + 15 = 90 \text{ mph for the express train}$$

7.

Formula	*Distance*	=	*Rate*	×	*Time*
Faster cyclist	$3(x+3)$		$x+3$		3 hr
Slower cyclist	$3x$		x		3 hr

d_1 = distance of faster cyclist
d_2 = distance of slower cyclist
$d_1 + d_2$ = total distance = 68 miles

Equation:

$$3(x+3)+3x=68$$
$$3x+9+3x=68$$
$$6x=59$$
$$x=9\tfrac{5}{6} \text{ mph for the slower cyclist}$$
$$x+3=12\tfrac{5}{6} \text{ mph for the faster cyclist}$$

8.

Formula	*Distance*	=	*Rate*	×	*Time*
1st steamboat	$6x$		x		6 hr
2nd steamboat	$6(x+9)$		$x+9$		6 hr

$d_1 + d_2$ = total distance covered by the 2 boats

Equation:

$$6x+6(x+9)=126$$
$$6x+6x+54=126$$
$$12x=72$$
$$x=6 \text{ mph for first boat}$$
$$x+9=15 \text{ mph for second boat}$$

9.

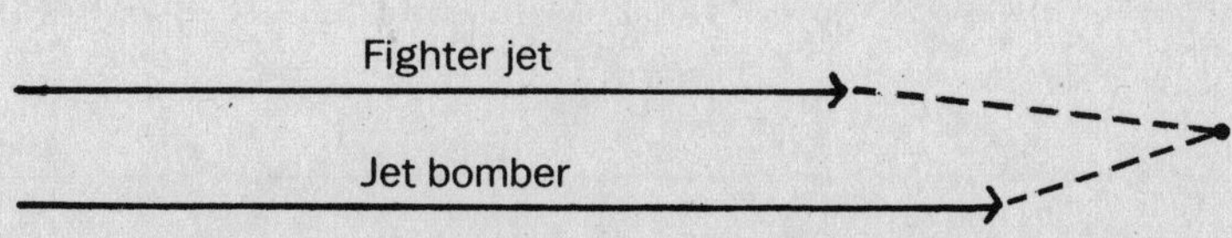

Formula	*Distance*	=	*Rate*	×	*Time*
Jet bomber	$500(x+.5)$		500		$x+.5$
Fighter jet	$600x$		600		x

Equation:

$$600x=500(x+.5)$$
$$600x=500x+250$$
$$600x-500x=250$$
$$100x=250$$
$$x=2\tfrac{1}{2} \text{ hr for fighter jet}$$
$$x+\tfrac{1}{2}=3 \text{ hr for jet bomber}$$

10.

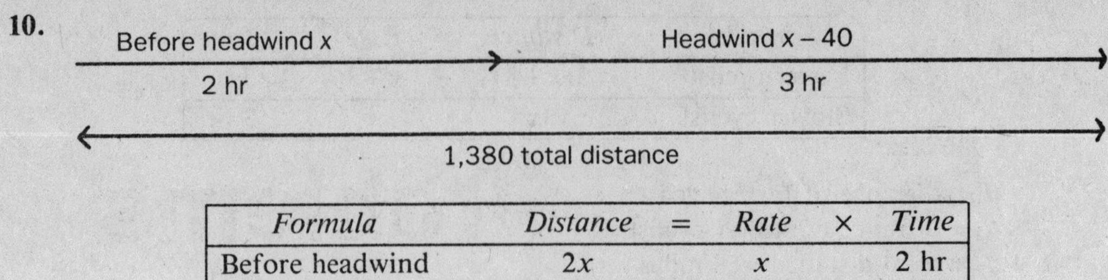

Formula	*Distance*	=	*Rate*	×	*Time*
Before headwind	$2x$		x		2 hr
With headwind	$3(x-40)$		$x-40$		3 hr

Equation: $2x+3(x-40)=1{,}380$ miles

$$2x+3x-120=1{,}380$$
$$5x-120=1{,}380$$
$$5x=1{,}380+120$$
$$5x=1{,}500$$

$x=300$ mph speed before flying into headwind

5 Coin Problems

When solving coin problems, all coins must be changed to cents.

Examples

$$1 \text{ nickel} = 5(1) = 5 \text{ cents}$$
$$2 \text{ nickels} = 5(2) = 10 \text{ cents}$$
$$5 \text{ nickels} = 5(5) = 25 \text{ cents}$$
$$x \text{ nickels} = 5(x) = 5x \text{ cents}$$
$$(x+2) \text{ nickels} = 5(x+2) = 5x+10 \text{ cents}$$
$$1 \text{ dime} = 10(1) = 10 \text{ cents}$$
$$5 \text{ dimes} = 10(5) = 50 \text{ cents}$$
$$x \text{ dimes} = 10(x) = 10x \text{ cents}$$
$$(x-2) \text{ dimes} = 10(x-2) = 10x-20 \text{ cents}$$
$$1 \text{ quarter} = 25(1) = 25 \text{ cents}$$
$$2 \text{ quarters} = 25(2) = 50 \text{ cents}$$
$$x \text{ quarters} = 25(x) = 25x \text{ cents}$$
$$1 \text{ half dollar} = 50(1) = 50 \text{ cents}$$
$$2 \text{ half dollars} = 50(2) = 100 \text{ cents}$$
$$x \text{ half dollars} = 50(x) = 50x \text{ cents}$$
$$3x \text{ half dollars} = 50(3x) = 150x \text{ cents}$$
$$(2x+5) \text{ half dollars} = 50(2x+5) = 100x+250 \text{ cents}$$

Since there are 100 cents in 1 dollar, then there are 540 cents in \$5.40. In \$11.45, there would be 1,145 cents.

STUDY PROBLEM 1

A collection of quarters and nickels contains twice as many quarters as nickels. If the value of the coins totals \$4.40, how many nickels and quarters are in the collection?

Let $x =$ number of nickels
$2x =$ number of quarters
$5(x) =$ value in cents of the number of nickels
$25(2x) =$ value in cents of the number of quarters
$\$4.40 = 440$ cents

Equation:

$$5x + 50x = 440 \text{ (all terms in cents)}$$
$$55x = 440$$
$$x = 8 \text{ (number of nickels)}$$
$$2x = 16 \text{ (number of quarters)}$$

Use the format in the example on page 23 to solve the following problems.

Practice Problems: Coins

(Check your answers with the solutions on pages 27–28.)

1. In a collection of coins that has a value of \$4.90, there are 20 more nickels than dimes. How many nickels and dimes are there in the collection?

Let x = number of __?__
__?__ = number of __?__
$10(x)$ = value in cents of __?__
$5(20+x)$ = value in cents of __?__
\$4.90 = __?__ cents

Equation: ________?________.
a. Number of dimes is __?__
b. Number of nickels is __?__

2. A student at Ellwood High has a collection of quarters and nickels amounting to \$18.70. If there are 20 more nickels than quarters, how many of each kind of coin does the student have?

Let x = number of __?__
__?__ = number of __?__
$25(x)$ = value in cents of __?__
$5(x+20)$ = value in cents of __?__
\$18.70 = __?__ cents

Equation: ________?________.
a. Number of quarters is __?__
b. Number of nickels is __?__

3. A woman has \$10.50 in change consisting of twice as many half dollars as quarters and five times as many dimes as quarters. How many half dollars, quarters, and dimes does she have?

Let x = number of __?__
$2x$ = number of __?__
$5x$ = number of __?__
$25(x)$ = value in cents of __?__
$50(2x)$ = value in cents of __?__
$10(5x)$ = value in cents of __?__
\$10.50 = __?__ cents

Equation: ________?________.
a. Number of quarters is __?__
b. Number of half-dollars is __?__
c. Number of dimes is __?__

4. Sylvia has \$11.25 in nickels, dimes, and quarters. She has 3 times as many nickels as dimes and 5 more quarters than dimes. How many of each kind of coin does she have?

Let x = number of __?__
$3x$ = number of __?__

(?) = number of ___?___
___?___ = value in cents of dimes
___?___ = value in cents of nickels
___?___ = value in cents of quarters
$11.25 = ___?___ cents

Equation: ___________?___________.

a. Number of dimes is ___?___
b. Number of nickels is ___?___
c. Number of quarters is ___?___

5. A news boy collected $68.50 on his paper route one week. The collection consisted of nickels, dimes, and quarters. If he had 70 more nickels than quarters and 20 more dimes than nickels, how many of each kind of coin did he have?

Let x = number of quarters
___?___ = number of nickels
___?___ = number of dimes
$25(x)$ = value in cents of quarters
___?___ = value in cents of nickels
___?___ = value in cents of dimes
$68.50 = ___?___ cents

Equation: ___________?___________.

a. Number of quarters is ___?___
b. Number of nickels is ___?___
c. Number of dimes is ___?___

Whenever the *total number* of coins, tickets, or stamps is given in a problem, follow the reasoning below.

If you have a stack of 22 coins made up of nickels and dimes and you are told that 12 of the coins are dimes, then $22-12=10$ nickels.

If you know that 7 of the coins are nickels, then $22-7=15$ dimes.

If you have x number of dimes, then $22-x$ = the number of nickels.

STUDY PROBLEM 2

A purse contained 58 coins consisting of dimes and nickels. If the total amount of these coins amounted to $4.80, how many of each kind of coin are in the purse?

Let x = number of dimes
$58-x$ = number of nickels
$10(x)$ = value in cents of x dimes
$5(58-x)$ = value in cents of $(58-x)$ nickels
$4.80 = 480 cents

Equation:

$$10x+5(58-x)=480$$
$$10x+290-5x=480$$
$$5x+290=480$$
$$5x=190$$
$$x=38 \text{ dimes}$$
$$58-x=20 \text{ nickels}$$

More Practice Problems

(Check your answers with the solutions on pages 28–29.)

6. Eighty-six coins made up of dimes and quarters amount to $14.60. How many of each of these coins are there?

Let x = number of quarters
(?) = number of dimes
? = value in cents of quarters
? = value in cents of dimes
$14.60 = ? cents

Equation: ________?________.
  a. Number of quarters is ?
  b. Number of dimes is ?

7. A piggy bank contained 110 coins including nickels, dimes, and quarters. If there are 20 more nickels than quarters, and the number of dimes is equal to the number of quarters, how many of each kind of coin are there if the total collection is $13.00?

Let x = number of ?
x = number of ?
(?) = number of nickels
$10x$ = value in cents of ?
$?x$ = value in cents of quarters
$5(20+x)$ = value in cents of ?
$13.00 = ? cents

Equation: ________?________.
  a. Number of dimes is ?
  b. Number of quarters is ?
  c. Number of nickels is ?

8. A high school play grossed $122.50. If there were twice as many student tickets sold as there were adult tickets, and if each student ticket sold for 50 cents and each adult ticket for 75 cents, how many student and how many adult tickets were sold for the play? (Follow the same procedure as in solving coin problems.)
  a. Adult tickets sold = ?
  b. Student tickets sold = ?

9. Ralph, who is president of the senior class at his high school, sold $67.50 worth of tickets for the music festival. If he sold 20 more student tickets than adult tickets, and the cost for student tickets was 75 cents and adult tickets $1.00, how many of each kind of ticket did he sell?
(Follow the same procedure as in solving coin problems.)
  a. Adult tickets sold = ?
  b. Student tickets sold = ?

10. Susie bought $5.61 worth of stamps. She bought the same number of 2- and 5-cent stamps and double that number of 22-cent stamps. How many of each kind of stamp did she buy?

Let $x =$ number of ___?___
___?___ = number of 5-cent stamps
___?___ = number of 22-cent stamps
___?___ = value in cents of 2-cent stamps
___?___ = value in cents of 5-cent stamps
___?___ = value in cents of 22-cent stamps
$5.61 = ___?___ cents

Equation: ___________?___________.

a. Number of 2-cent stamps is ___?___
b. Number of 5-cent stamps is ___?___
c. Number of 22-cent stamps is ___?___

Answers to Practice Problems: Coins

1.

$x =$ number of dimes
$x+20 =$ number of nickels
$10(x) =$ value in cents of dimes
$5(20+x) =$ value in cents of nickels
$4.90 = 490 cents

Equation:

$$10x + 5(20+x) = 490$$
$$10x + 100 + 5x = 490$$
$$15x = 490 - 100$$
$$15x = 390$$
$$x = 26 \text{ dimes}$$
$$20 + x = 46 \text{ nickels}$$

2.

$x =$ number of quarters
$x+20 =$ number of nickels
$25(x) =$ value in cents of quarters
$5(x+20) =$ value in cents of nickels
$18.70 = 1,870 cents

Equation:

$$25x + 5(x+20) = 1{,}870$$
$$25x + 5x + 100 = 1{,}870$$
$$30x + 100 = 1{,}870$$
$$30x = 1{,}870 - 100$$
$$30x = 1{,}770$$
$$x = 59 \text{ quarters}$$
$$x + 20 = 79 \text{ nickels}$$

3.

$x =$ number of quarters
$2x =$ number of half-dollars
$5x =$ number of dimes
$25(x) =$ value in cents of quarters
$50(2x) =$ value in cents of half-dollars
$10(5x) =$ value in cents of dimes
$10.50 = 1,050 cents

Equation:

$$25x+100x+50x=1{,}050$$
$$175x=1{,}050$$
$$x=\underline{6\text{ quarters}}$$
$$2x=\underline{12\text{ half-dollars}}$$
$$5x=\underline{30\text{ dimes}}$$

4.

$$x=\text{number of }\underline{\text{dimes}}$$
$$3x=\text{number of }\underline{\text{nickels}}$$
$$\underline{x+5}=\text{number of }\underline{\text{quarters}}$$
$$\underline{10(x)}=\text{value in cents of dimes}$$
$$\underline{5(3x)}=\text{value in cents of nickels}$$
$$\underline{25(x+5)}=\text{value in cents of quarters}$$
$$\$11.25=\underline{1{,}125}\text{ cents}$$

Equation:

$$10(x)+5(3x)+25(x+5)=1{,}125$$
$$10x+15x+25x+125=1{,}125$$
$$50x+125=1{,}125$$
$$50x=1{,}125-125$$
$$50x=1{,}000$$
$$x=\underline{20\text{ dimes}}$$
$$3x=\underline{60\text{ nickels}}$$
$$(x+5)=\underline{25\text{ quarters}}$$

5.

$$\text{Let }x=\text{number of quarters}$$
$$\underline{70+x}=\text{number of nickels}$$
$$\underline{90+x}=\text{number of dimes}$$
$$25(x)=\text{value in cents of quarters}$$
$$\underline{5(70+x)}=\text{value in cents of nickels}$$
$$\underline{10(90+x)}=\text{value in cents of dimes}$$
$$\$68.50=\underline{6{,}850}\text{ cents}$$

Equation:

$$25x+5(70+x)+10(90+x)=6{,}850$$
$$25x+350+5x+900+10x=6{,}850$$
$$40x=6{,}850-1{,}250$$
$$40x=5{,}600$$
$$x=\underline{140\text{ quarters}}$$
$$70+x=\underline{210\text{ nickels}}$$
$$90+x=\underline{230\text{ dimes}}$$

6.

$$\text{Let }x=\text{number of quarters}$$
$$\underline{86-x}=\text{number of dimes}$$
$$\underline{25(x)}=\text{value in cents of quarters}$$
$$\underline{10(86-x)}=\text{value in cents of dimes}$$
$$\$14.60=\underline{1{,}460}\text{ cents}$$

Equation:

$$25(x)+10(86-x)=1{,}460$$
$$25x+860-10x=1{,}460$$
$$15x=1{,}460-860$$
$$15x=600$$
$$x=\underline{40\text{ quarters}}$$
$$86-x=\underline{46\text{ dimes}}$$

7.

$$x=\text{number of }\underline{\text{quarters}}$$
$$x=\text{number of }\underline{\text{dimes}}$$

$\underline{20+x}$ = number of nickels
$10x$ = value in cents of $\underline{\text{dimes}}$
$\underline{25x}$ = value in cents of quarters
$5(20+x)$ = value in cents of $\underline{\text{nickels}}$
$13.00 = $\underline{1,300}$ cents

Equation:

$$25x+10x+5(20+x)=1{,}300$$
$$25x+10x+100+5x=1{,}300$$
$$40x=1{,}300-100$$
$$40x=1{,}200$$
$$x=\underline{30\text{ dimes}}$$
$$x=\underline{30\text{ quarters}}$$
$$20+x=\underline{50\text{ nickels}}$$

8. Let x = number of adult tickets sold
$2x$ = number of student tickets sold
$75(x)$ = value in cents of the adult tickets sold
$50(2x)$ = value in cents of student tickets sold
$122.50 = 12,250 cents

Equation:

$$75x+100x=12{,}250$$
$$175x=12{,}250$$
$$x=\underline{70}\text{ adult tickets sold}$$
$$2x=\underline{140}\text{ student tickets sold}$$

9. Let x = number of adult tickets sold
$x+20$ = number of student tickets sold
$100(x)$ = value in cents of adult tickets sold
$75(x+20)$ = value in cents of student tickets sold
$67.50 = 6,750 cents

Equation:

$$100x+75(x+20)=6{,}750$$
$$100x+75x+1{,}500=6{,}750$$
$$175x=6{,}750-1{,}500$$
$$175x=5{,}250$$
$$x=\underline{30}\text{ adult tickets sold}$$
$$x+20=\underline{50}\text{ student tickets sold}$$

10. Let x = number of $\underline{\text{2-cent stamps}}$
$\underline{x}$ = number of 5-cent stamps
$\underline{2x}$ = number of 22-cent stamps
$\underline{2x}$ = value in cents of 2-cent stamps
$\underline{5x}$ = value in cents of 5-cent stamps
$\underline{22(2x)}$ = value in cents of 22-cent stamps
$5.61 = $\underline{561}$ cents

Equation:

$$2x+5x+44x=561$$
$$51x=561$$
$$x=11\text{ 2-cent stamps}$$
$$x=11\text{ 5-cent stamps}$$
$$2x=22\text{ 22-cent stamps}$$

6 Mixture Problems

REVIEW OF PERCENT

Without adequate knowledge of percents, mixture problems will be impossible for you to solve. Take the review test below without looking at the answers on the next page. Study the answers that you missed. If you do not understand, go back and review the test before going on.

Part I

Percent means hundredths. Change the following percents to decimal fractions.
Example: 5% is .05

1. 3%	6. 37%	11. 5½%	16. 275%	21. 6.8%
2. 4%	7. 44%	12. 6½%	17. 8¼%	22. 12.8%
3. 8%	8. 59%	13. 8½%	18. 9¾%	23. 127.3%
4. 11%	9. 46%	14. 12½%	19. 10¾%	24. 2.8%
5. 22%	10. 88%	15. 155%	20. 112½%	25. .5%

Part II

Change the percents below to decimals before multiplying.

1. 5% of 250 = ?	6. 12¼% of 800 = ?
2. 7% of 336 = ?	7. 6.5% of $44,500 = ?
3. 32% of 12,500 = ?	8. 300% of 50 = ?
4. 6% of x = ?	9. 8½% of $6,180 = ?
5. 6% of $(10{,}000 - x)$ = ?	10. .8% of 42,000 = ?

Check your answers before going on.

ANSWERS TO REVIEW OF PERCENT

Part I

1. .03
2. .04
3. .08
4. .11
5. .22
6. .37
7. .44
8. .59
9. .46
10. .88
11. .05½ or .055
12. .06½ or .065
13. .08½ or .085
14. .12½ or .125
15. 1.55
16. 2.75
17. .08¼ or .0825
18. .09¾ or .0975
19. .10¾ or .1075
20. 1.12½ or 1.125
21. .068
22. .128
23. 1.273
24. .028
25. .005

Part II

1. 12.50
2. 23.52
3. 4,000.00
4. $.06x$
5. $600 - .06x$
6. 98.00
7. $2,892.50
8. 150.00
9. $525.30
10. $336.00

When solving mixture problems, there will be 3 different percentages of 3 different costs per pound of candy, nuts, fruits, or other product. When the problem involves chemicals or other liquids, a liquid measurement such as ounces, grams, cubic centimeters, liters, quarts, gallons, or pints will be given. We will use the model below to help us find the solution to mixture problems.

Model:

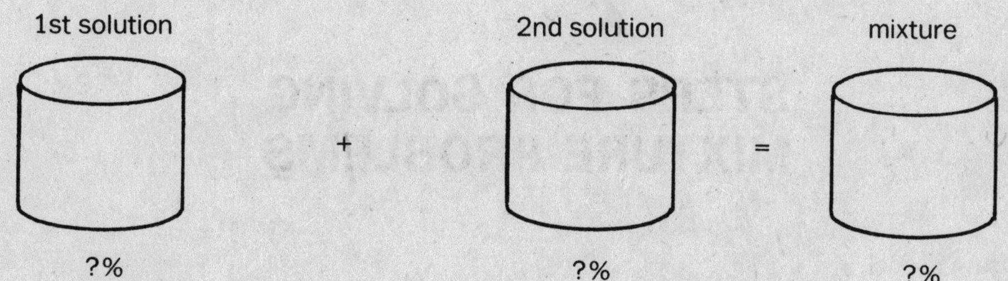

STUDY PROBLEM 1

How many ounces of pure alcohol must be added to a 42-oz solution containing 10% alcohol to make a solution containing 40% alcohol?

Model:

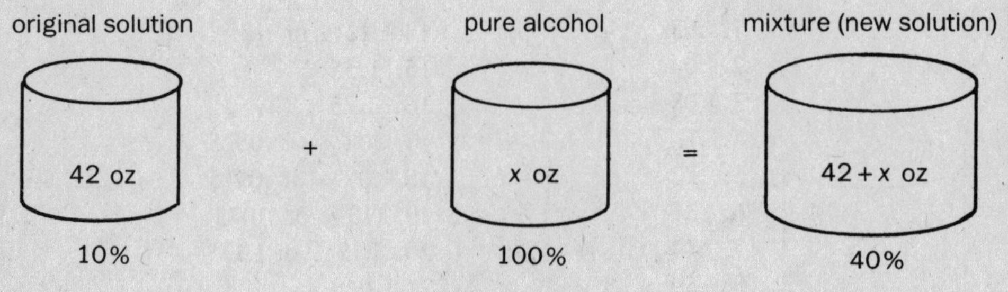

original solution		pure alcohol		mixture (new solution)
42 oz	+	x oz	=	42 + x oz
10%		100%		40%

Change each percent to a decimal fraction as follows:

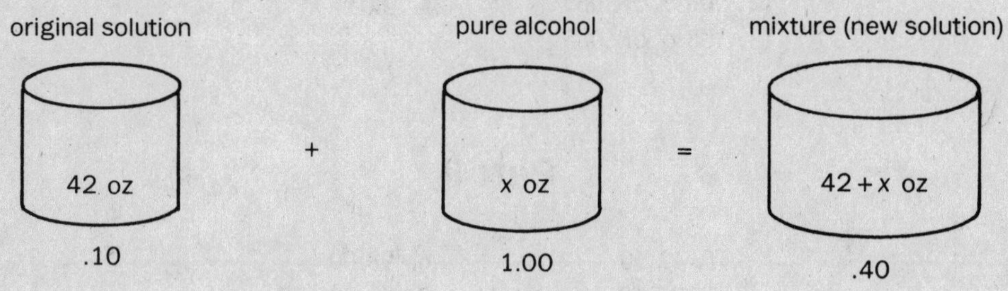

original solution		pure alcohol		mixture (new solution)
42 oz	+	x oz	=	42 + x oz
.10		1.00		.40

Multiply each decimal fraction by the contents in each container:

Equation: $.10(42) + 1.00(x) = .40(42 + x)$

$$4.20 + 1.00x = 16.80 + .40x$$

$$100x - 40x = 1{,}680 - 420$$

$$60x = 1{,}260$$

$x = 21$ oz of pure alcohol must be added.

The new mixture will contain 63 oz.

STEPS FOR SOLVING MIXTURE PROBLEMS

1. Draw model.
2. Place percents outside containers.
3. Change percents to decimals (hundredths).

4. Multiply each decimal fraction by the contents in each container.
5. Write equation.
6. Solve equation.

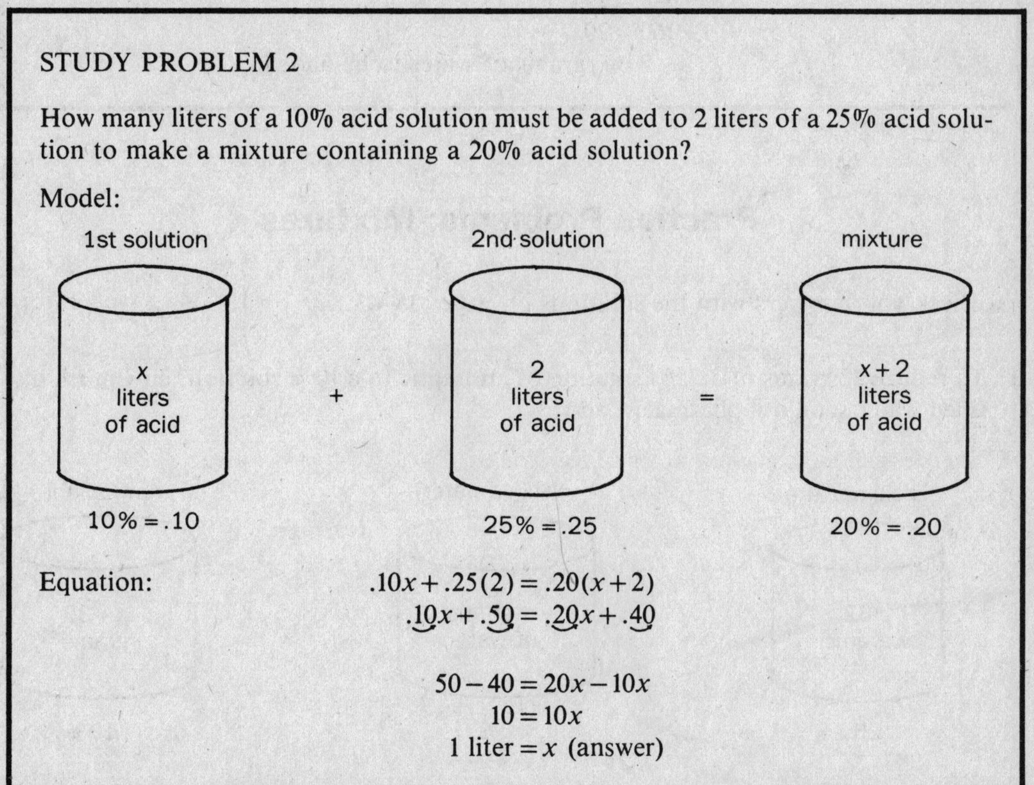

STUDY PROBLEM 2

How many liters of a 10% acid solution must be added to 2 liters of a 25% acid solution to make a mixture containing a 20% acid solution?

Model:

Equation:

$$.10x + .25(2) = .20(x+2)$$
$$.10x + .50 = .20x + .40$$

$$50 - 40 = 20x - 10x$$
$$10 = 10x$$
$$1 \text{ liter} = x \text{ (answer)}$$

Note: When using water to reduce the percentage of the solution, use 0% on the outside of the container that represents the water.

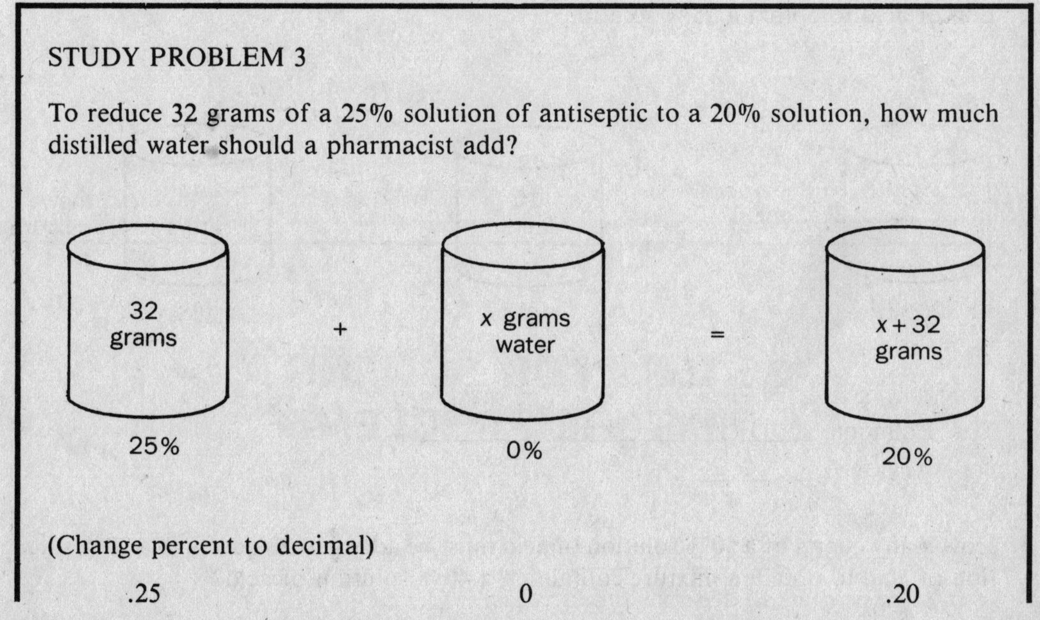

STUDY PROBLEM 3

To reduce 32 grams of a 25% solution of antiseptic to a 20% solution, how much distilled water should a pharmacist add?

(Change percent to decimal)

.25 0 .20

Equation: $.25(32)+0=.20(x+32)$

$8.00+0=.20x+6.40$

$800-640=20x$

$160=20x$

$8=x$ grams of water to be added

Practice Problems: Mixtures

(Check your answers with the solutions on pages 38–43.)

1. To reduce 32 grams of a 25% solution of antiseptic to a 10% solution, how much distilled water should a pharmacist add?

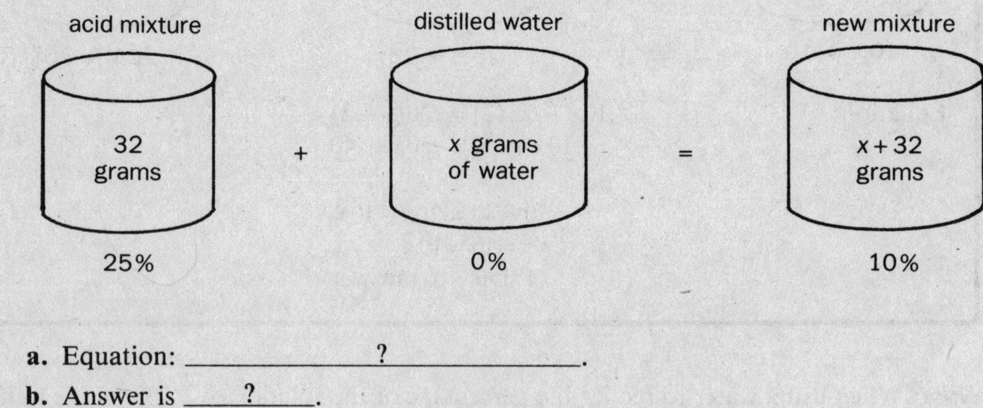

a. Equation: ________?________.

b. Answer is ___?___.

2. How many liters of a 20% solution of acid should be added to 10 liters of a 30% solution of acid to obtain a 25% solution?

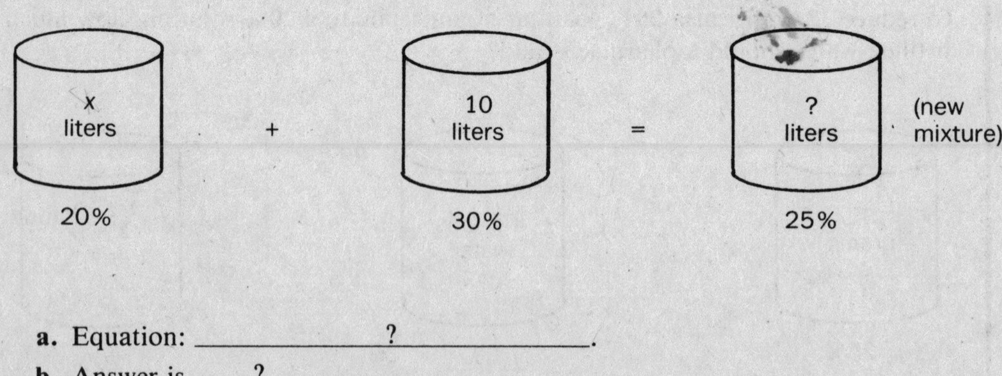

a. Equation: ________?________.

b. Answer is ___?___.

3. How many quarts of a 50% solution of acid must be added to 20 quarts of a 20% solution of acid to obtain a mixture containing a 40% solution of acid?

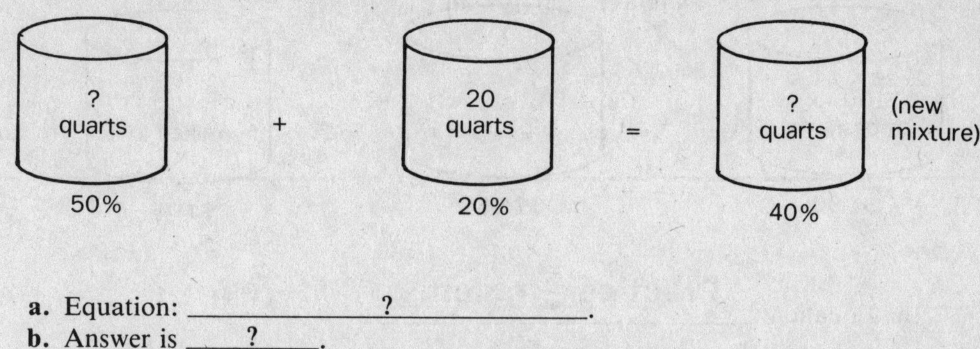

a. Equation: ________?________.
b. Answer is ___?___.

4. A chemist has a solution of 50% pure acid and another of 80% pure acid. How many gallons of each will make 600 gallons of 72% pure acid?

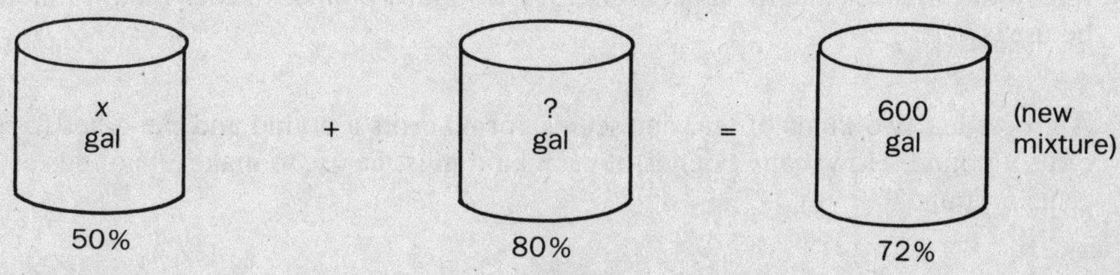

a. Equation: ________?________.
b. Answer is __?__ gallons for 50% pure acid.
c. Answer is __?__ gallons for 80% pure acid.

5. How much alcohol must be added to 1 quart of tincture of arnica containing 20% arnica to reduce it to a 10% arnica solution? *No Help This Time!*

6. How many quarts of peanut oil worth 19 cents a quart must be mixed with 100 quarts worth 25 cents a quart to produce a mixture which will sell for 24 cents?

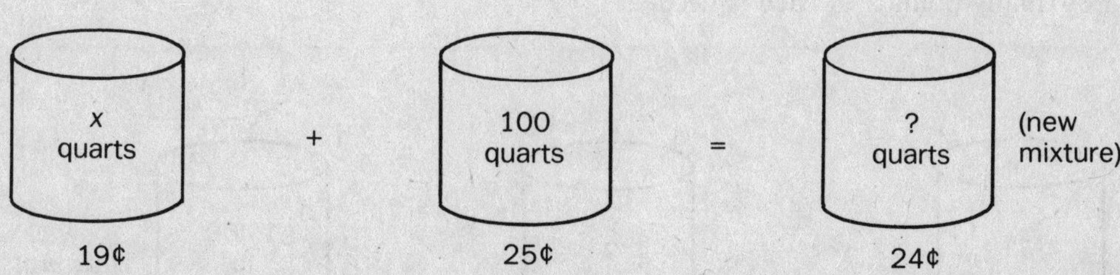

7. A storekeeper wishes to sell 100 pounds of mixed nuts at \$1.75 a pound. He mixes peanuts worth \$1.65 a pound with cashews worth \$1.90 a pound. How many pounds of each does he use?

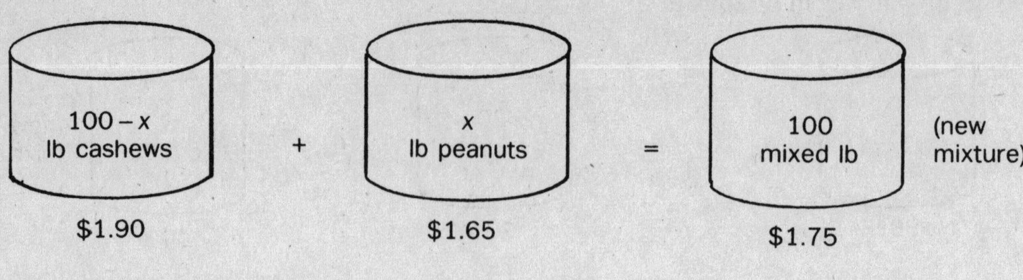

a. Equation: ______?______.

b. __?__ lb at $1.65.

c. __?__ lb at $1.90.

8. A merchant wishes to blend 200 pounds of coffee worth $3.20 a pound from two mixtures—one at $3.00 and the other at $3.25. How many pounds of each mixture should he use?

9. A grocer has two kinds of tea, one selling for 80 cents a pound and the other for 60 cents a pound. How many pounds of each kind must he use to make 50 pounds at 74 cents a pound?

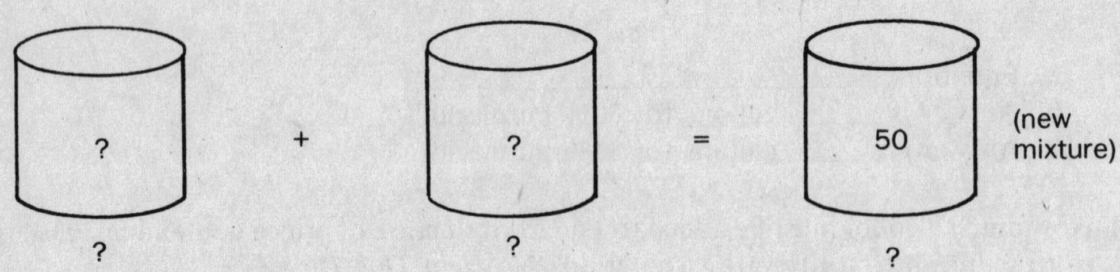

10. A dairy wishes to mix 1,000 pounds of milk containing 8% butterfat. If the mixture is to be made from milk containing 5% butterfat and cream containing 20% butterfat, how many pounds of each is needed?

a. Equation: ______?______.

b. Answer in pounds at 5% is __?__.

c. Answer in pounds at 20% is __?__.

11. How many ounces of an alloy containing 30% gold on the world market must be mixed with an alloy containing 5% gold to obtain 25 ounces of an alloy containing 20% gold?

12. A pharmacist has a 12% solution and a 20% solution of boric acid. How much of each must he use to make 80 grams of a 15% solution?

13. A druggist has an 18% solution of argyol. How much of this solution and how much water must be mixed together to make 10 liters of a 12% solution?

14. A grocer wishes to blend two different coffee brands. He wants to make a blend of 480 pounds to sell at $2.68 a pound. If he uses a blend of coffee worth $2.50 a pound with another blend worth $2.80 a pound, how many pounds of each does he use?

15. How many ounces of iodine worth 30 cents an ounce must be mixed with 50 ounces of iodine worth 18 cents an ounce so that the mixture can be sold for 20 cents an ounce?

16. How many gallons of rocket fuel containing 80% liquid oxygen must be mixed with 250 gallons containing 40% liquid oxygen to produce a mixture containing 70% oxygen?

17. The Patronete winery's tastiest wine must contain a 12% alcohol content. How many gallons of wine with a 9% alcohol content must be mixed with 3,000 gallons of wine with a 15% alcohol content in order to achieve the desired 12% alcohol content?

18. The Davidson winery is noted for its Regal champagne. It takes a blending of 1,000 bushels of a mixture of two varieties of grapes to achieve the Regal specification of 12% alcohol content for its champagne. One of the two varieties of grape contains a 10% alcohol content and the other contains a 15% alcohol content. How many bushels of each variety are needed?

Note: In some mixture problems you not only add something to the mixture, but you also drain off some of the mixture. Four containers will be used in the following format.

STUDY PROBLEM 4

A 7-quart truck radiator was assumed to contain an 80% rating in alcohol, but after checking, it was found to contain a rating of only 60% alcohol. How much 90% alcohol must be added and how much 60% alcohol must be drained off to achieve an 80% rating?

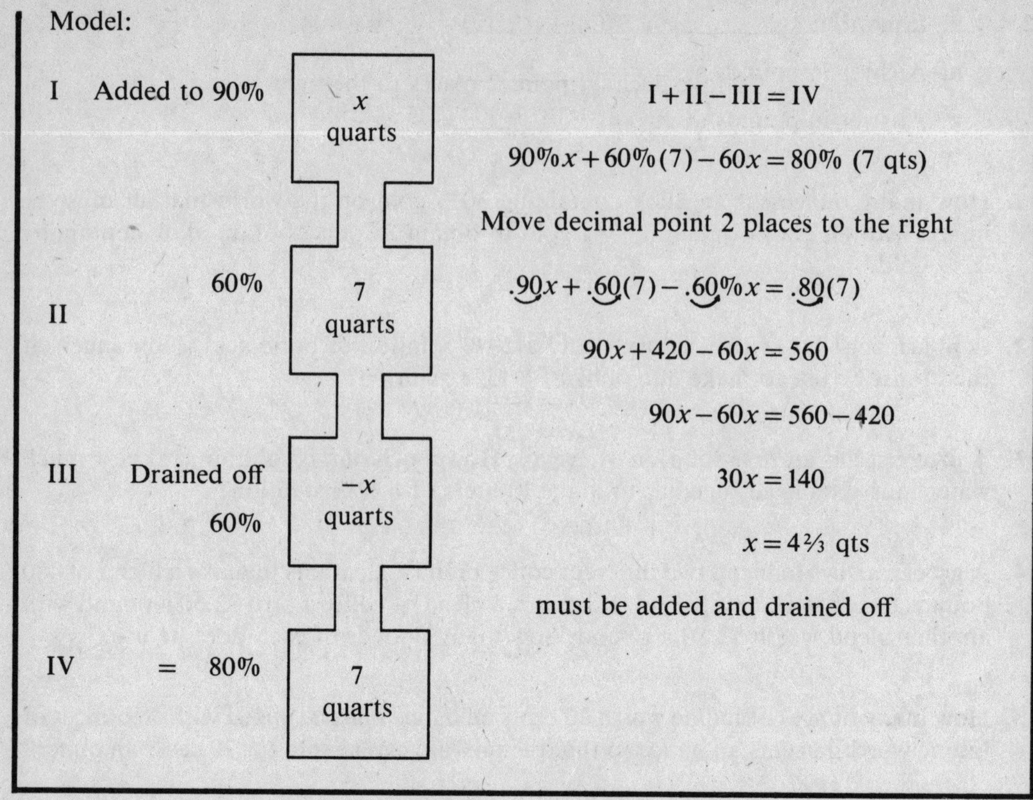

More Practice Problems

(Check your answers with the solutions on page 44.)

19. The capacity of a car radiator is 20 quarts. If it's full of a 55% antifreeze solution, how many quarts must be drained off and replaced by a 80% solution to give 20 quarts of a 70% solution?

20. The capacity of a car radiator is 18 quarts. If it's full of a 20% antifreeze solution, how many quarts must be drained off and replaced by a 100% solution to give 18 quarts of a 39% solution?

Answers to Practice Problems: Mixtures

1. a.

$$.25(32) + 0\%(x) = .10(x + 32)$$

Move decimal point 2 places to the right

$$25(32) + 0x = 10(x + 32)$$
$$800 + 0 = 10x + 320$$
$$800 - 320 = 10x$$
$$480 = 10x$$

b. 48 grams of water $= x$ to be added

2. a. $$.20(x) + .30(10) = .25(x + 10)$$

Move decimal point 2 places to the right

$$20x + 300 = 25x + 250$$
$$300 - 250 = 25x - 20x$$
$$50 = 5x$$

b. 10 liters of 20% acid $= x$

3. a. $$.50x + .20(20) = .40(x + 20)$$

Move decimal point 2 places to the right

$$50x + 400 = 40x + 800$$
$$50x - 40x = 800 - 400$$
$$10x = 400$$

b. $x = 40$ quarts to be added

4. a. $$.50(x) + .80(600 - x) = .72(600)$$

Move decimal point 2 places to the right

$$50x + 48{,}000 - 80x = 43{,}200$$
$$50x - 80x = -48{,}000 + 43{,}200$$
$$-30x = -4{,}800$$

b. $x = 160$ gal @ 50%

c. $600 - x = 440$ gal @ 80%

5.

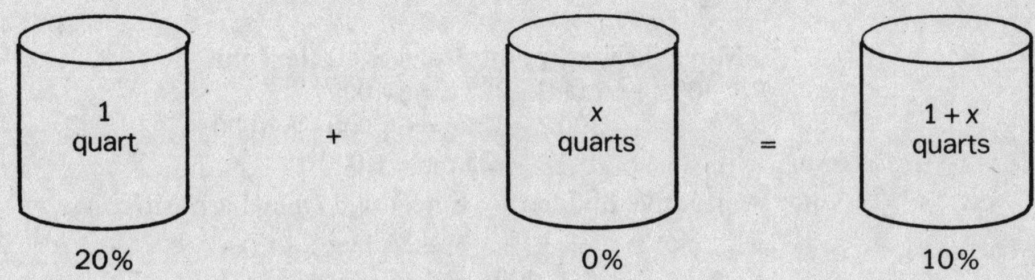

Equation: $$.20(1) + 0(x) = .10(1 + x)$$

Move decimal point 2 places to the right

$$20 + 0 = 10 + 10x$$
$$20 - 10 = 10x$$
$$10 = 10x$$
$$1 = x$$

Therefore, 1 quart of alcohol is needed.

6. Equation: $$.19(x) + .25(100) = .24(x + 100)$$

Move decimal point 2 places to the right

$$19x + 2{,}500 = 24x + 2{,}400$$
$$2{,}500 - 2{,}400 = 24x - 19x$$
$$100 = 5x$$
$$20 = x$$

Therefore, 20 quarts of 19-cent peanut oil are needed.

7. $x =$ lb of peanuts
$(100 - x) =$ lb of cashews

a. $1.90(100 - x) + 1.65(x) = 1.75(100)$

Move decimal point 2 places to the right

$$19{,}000 + 190x + 165x = 17{,}500$$
$$-25x = 17{,}500 - 19{,}000$$
$$-25x = -1{,}500$$
$$x = 60 \text{ lb peanuts @ \$1.65}$$
$$100 - x = 40 \text{ lb cashews @ \$1.90}$$

(*Note:* A negative divided by a negative equals a positive.)

b. $x = 60$ lb of peanuts

c. $100 - x = 40$ lb of cashews

8.

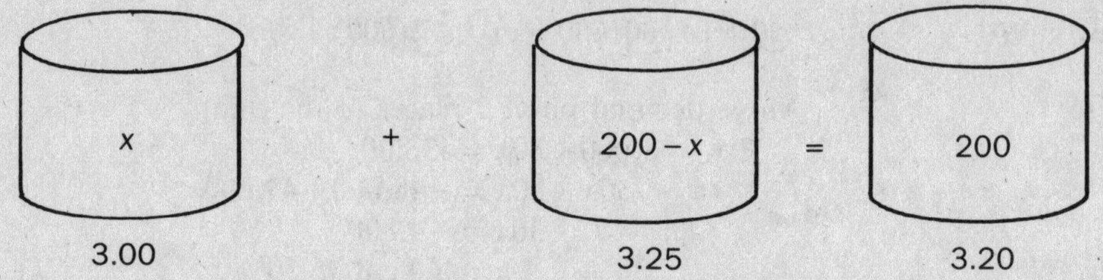

Equation: $3.00(x) + 3.25(200 - x) = 3.20(200)$

Move decimal point 2 places to the right

$$300x + 65{,}000 - 325x = 64{,}000$$
$$300x - 325x = 64{,}000 - 65{,}000$$
$$-25x = -1{,}000$$

(*Note:* A negative divided by a negative equals a positive.)

$$x = 40 \text{ lb @ \$3.00}$$
$$200 - x = 160 \text{ lb @ \$3.25}$$

9.

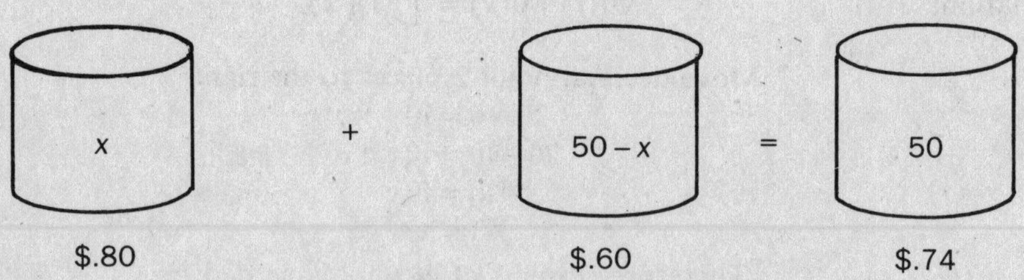

Equation: $.80(x) + .60(50 - x) = .74(50)$

Move decimal point 2 places to the right

$$80x + 3{,}000 - 60x = 3{,}700$$
$$20x = 3{,}700 - 3{,}000$$
$$20x = 700$$
$$x = 35 \text{ lb @ \$.80}$$
$$50 - x = 15 \text{ lb @ \$.60}$$

10.

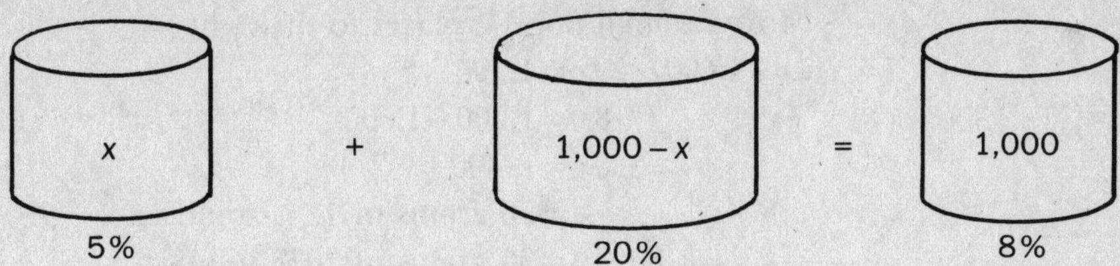

a. $.05(x) + .20(1{,}000 - x) = .08(1{,}000)$

Move decimal point 2 places to the right

$5x + 20{,}000 - 20x = 8{,}000$

$5x - 20x = 8{,}000 - 20{,}000$

$-15x = -12{,}000$

b. $x = 800$ lb @ 5%

c. $1000 - x = 200$ lb @ 20%

11.

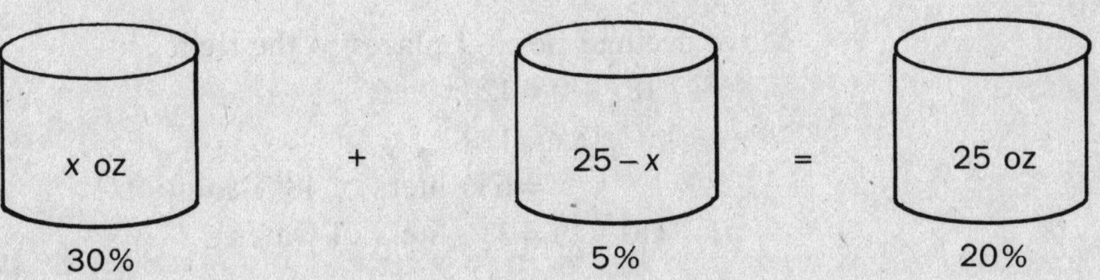

Equation: $.30(x) + .05(25 - x) = .20(25)$

Move decimal point 2 places to the right

$30x + 125 - 5x = 500$

$30x - 5x = 500 - 125$

$25x = 375$

$x = 15$ oz of a 30% alloy

$25 - x = 10$ oz of a 5% alloy

12.

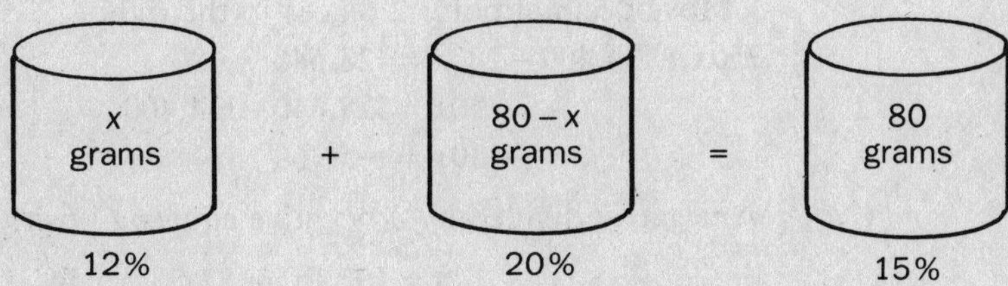

Equation: $.12(x) + .20(80 - x) = .15(80)$

Move decimal point 2 places to the right

$$12x + 1{,}600 - 20x = 1{,}200$$
$$-8x = 1{,}200 - 1{,}600$$
$$-8x = -400$$
$$x = 50 \text{ grams of } 12\% \text{ solution}$$
$$80 - x = 30 \text{ grams of } 20\% \text{ solution}$$

13.

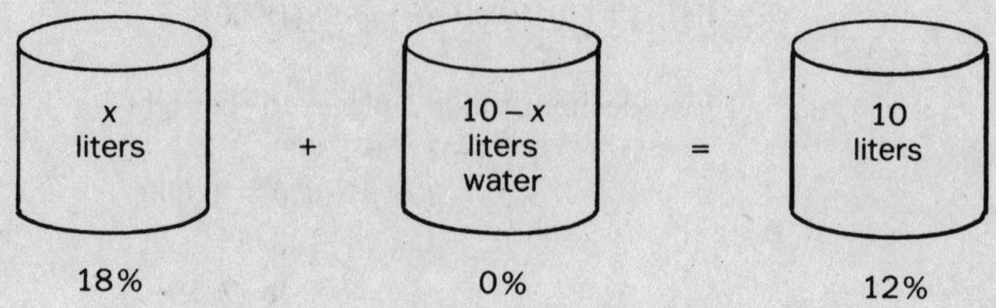

18% 0% 12%

Equation: $.18(x) + 0\%(10 - x) = .12(10)$

Move decimal point 2 places to the right

$$18x + 0 = 120$$
$$18x = 120$$
$$x = 6\tfrac{2}{3} \text{ liters of } 18\% \text{ solution}$$
$$(10 - x) = 3\tfrac{1}{3} \text{ liters of water}$$

14.

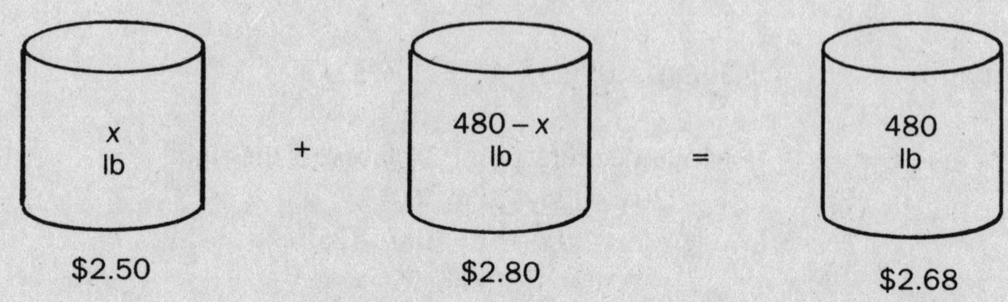

\$2.50 \$2.80 \$2.68

Equation: $2.50(x) + 2.80(480 - x) = 2.68(480)$

Move decimal point 2 places to the right

$$250x + 134{,}400 - 280x = 128{,}640$$
$$-30x = 128{,}640 - 134{,}400$$
$$-30x = -5{,}760$$

(*Note:* A negative divided by a negative equals a positive.)

$$x = 192 \text{ lb @ \$2.50 per lb}$$
$$480 - x = 288 \text{ lb @ \$2.80 per lb}$$

15.

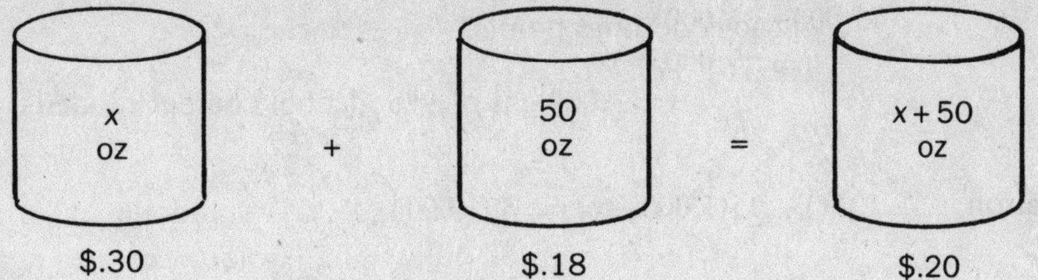

Equation: $.30(x) + .18(50) = .20(x + 50)$

Move decimal point 2 places to the right

$$30x + 900 = 20x + 1{,}000$$
$$30x - 20x = 1{,}000 - 900$$
$$10x = 100$$
$$x = 10 \text{ oz @ 30 cents}$$

16.

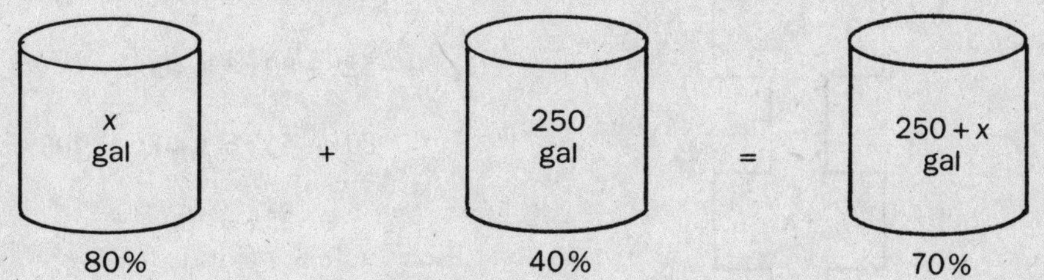

Equation: $.80(x) + .40(250) = .70(250 + x)$

Move decimal point 2 places to the right

$$80x + 10{,}000 = 17{,}500 + 70x$$
$$80x - 70x = 17{,}500 - 10{,}000$$
$$10x = 7{,}500$$
$$x = 750 \text{ gal of 80\% liquid oxygen needed}$$

17.

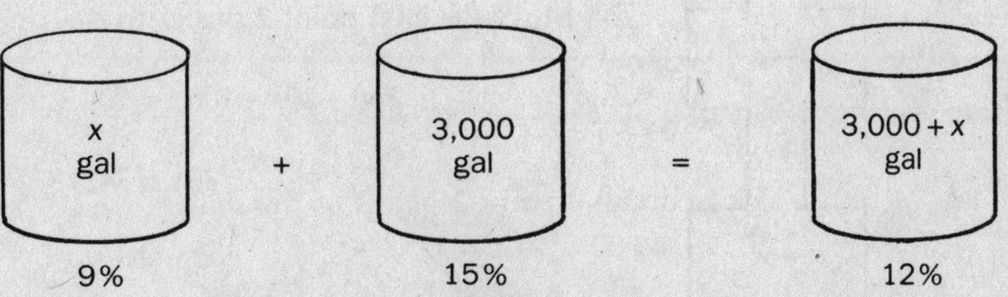

Equation: $.09(x) + .15(3{,}000) = .12(3{,}000 + x)$

Move decimal point 2 places to the right

$$9x + 45{,}000 = 36{,}000 + 12x$$
$$45{,}000 - 36{,}000 = 12x - 9x$$
$$9{,}000 = 3x$$
$$x = 3{,}000 \text{ gal of 9\% alcohol content needed}$$

18. Equation: $.10(x) + .15(1{,}000 - x) = .12(1{,}000)$

Move decimal point 2 places to the right

$$10x + 15{,}000 - 15x = 12{,}000$$
$$-5x = -3{,}000$$
$$x = 600 \text{ bushels of the 10\% variety}$$
$$1{,}000 - x = 400 \text{ bushels of the 15\% variety}$$

19.

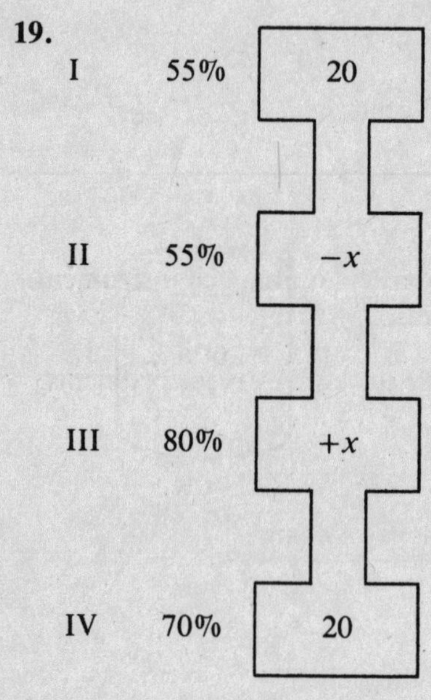

$$\text{I} - \text{II} + \text{III} = \text{IV}$$
$$.55(20) - .55(x) + .80(x) = .70(20)$$

Move decimal point 2 places to the right

$$1{,}100 - 55x + 80x = 1{,}400$$
$$80x - 55x = 1{,}400 - 1{,}100$$
$$25x = 300$$
$$x = 12 \text{ quarts}$$

must be drained off and replaced by an 80% solution

20.

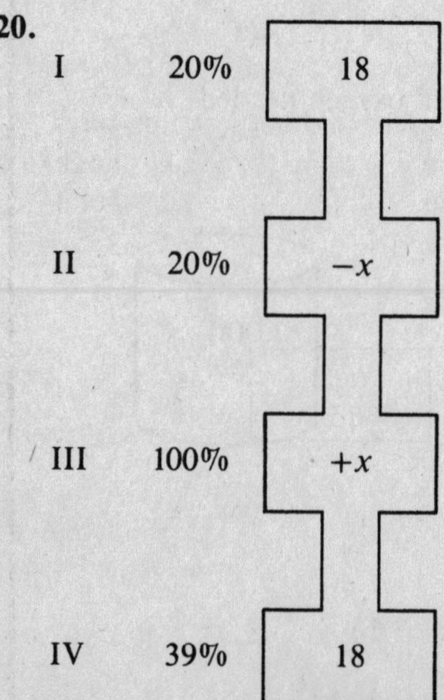

$$\text{I} - \text{II} + \text{III} = \text{IV}$$
$$.20(18) - .20(x) + 1.00(x) = .39(18)$$

Move decimal point 2 places to the right

$$360 - 20x + 100x = 702$$
$$80x = 342$$
$$x = 4\tfrac{11}{40}$$

quarts of a 20% solution must be drained off and replaced by a 100% solution

7 Investment Problems

Definition of terms used in the investment business:

1. Financial instruments:
   a. Common stocks
   b. Corporate bonds
   c. Real estate mortgages
   d. Savings accounts
   e. U.S. Treasury Notes
2. Yield: Interest or dividends earned or income from a particular financial instrument. All the instruments in item 1 earn interest or dividends.

This model represents any financial instrument. Use it to solve investment problems.

Model:

Dollars invested	+	Dollars invested	=	Yield (Money earned)
Rate of interest		Rate of interest		

STUDY PROBLEM 1

A man wishes to invest \$10,000. He places a certain amount in a bank which pays 6% interest and the remainder in a stock which yields a 7% dividend. How much money did he invest in the bank and how much in stock if his total earnings for 1 year are \$665.00?

Bank		Stock		Annual
Dollars x	+	Dollars $\$10{,}000 - x$	=	Yield \$665.00
$6\% = .06$		$7\% = .07$		

Equation: $.06x + .07(\$10{,}000 - x) = \$665.00$

Move decimal point 2 places to the right

$$6x + \$70{,}000 - 7x = \$66{,}500$$
$$-x = \$66{,}500 - \$70{,}000$$

$$-x = -\$3{,}500^*$$
$$x = \$3{,}500 \text{ invested in the bank}$$
$$\$10{,}000 - x = \$6{,}500 \text{ invested in stocks}$$

*Reminder: A negative divided by a negative becomes a positive.

STUDY PROBLEM 2

Mr. Kim, a diamond dealer, invested part of his $10,000 in a U.S. Treasury Note at 8½% and the remainder in a corporate bond at 10½%. If the income from both of these investments yielded $970.00, how much did he invest at each rate?

Treasury Note		Corporate Bond		Annual Yield
x	+	$10,000 − x	=	$970.00
8½%		10½%		

Equation: $.08\tfrac{1}{2}(x) + .10\tfrac{1}{2}(10{,}000 - x) = 970.00$

Move decimal point 2 places to the right

$$8\tfrac{1}{2}x + 105{,}000 - 10\tfrac{1}{2}x = 97{,}000$$
$$-2x = -8{,}000$$
$$x = \$4{,}000 \text{ invested at } 8\tfrac{1}{2}\%$$
$$10{,}000 - x = \$6{,}000 \text{ invested at } 10\tfrac{1}{2}\%$$

STUDY PROBLEM 3

The income from the first investment is $500.00 more than the income from the second investment. If $1,000 more is invested in the first investment which pays 9% than in the second investment paying 8%, how much was invested in each?

1st Investment		2nd Investment		Annual Yield
x + $1,000	−	x	=	$500.00
9%		8%		

Equation: $.09(x + 1{,}000) - .08(x) = 500.00$

Move decimal point 2 places to the right

$$9x + 9{,}000 - 8x = 50{,}000$$
$$9x - 8x = 50{,}000 - 9{,}000$$
$$x = \$41{,}000 \text{ invested at } 8\%$$
$$x + 1{,}000 = \$42{,}000 \text{ invested at } 9\%$$

Now solve these problems. Check your answers with those on pages 48–51.

Practice Problems: Investment

1. On her sixteenth birthday Jane received \$1,000 from her grandfather. If she invested the money in 2 banks, one paying 6% and the other 7%, how much did she invest in each bank if her annual interest was \$65.00?

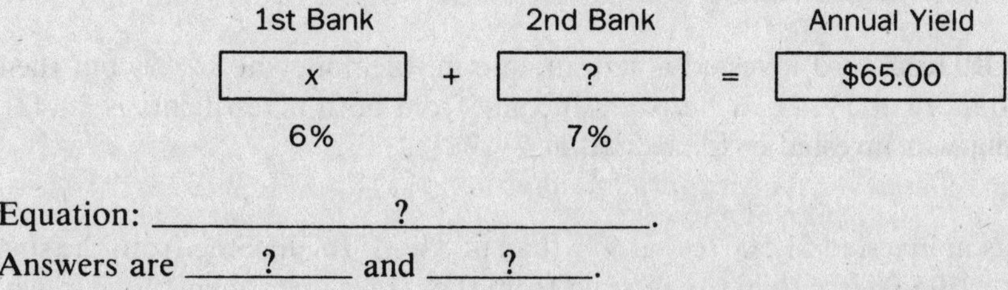

Equation: ______?______.
Answers are ___?___ and ___?___.

2. Part of \$12,000 is invested in 9% bonds and the remainder in common stocks paying 8%. If the yearly income from both investments is \$1,040.00, how much was invested in bonds and how much was invested in stocks?

3. A newly married couple has \$4,200 to invest. If they invest in one real estate mortgage paying 9½% and another paying 9%, how much did they invest in each mortgage if their yearly earnings are \$392.00?

4. Ralph's uncle invested \$1,000 more in a 6% investment than in a 5% investment. If his yearly income from both investments is \$1,710.00, how much did he invest at each rate?

5. Mrs. Walsh has \$14,000 to invest. She chooses two investments, one paying 8½% and one paying 7½%. If her yearly earnings from both investments is \$1,152.00, what did she invest at 8½%?

6. Mr. Camacho invested twice as much in a savings account at 6½% as he invested in a U.S. Treasury Note at 7%. If his yearly income from both investments is \$800.00, how much has he invested at 6½% and how much at 7%?

Savings Account		Treasury Note		Annual Yield
?	+	x	=	\$800.00
6½%		7%		

Equation: ______?______.
Amount invested at 6½% is ___?___.
Amount invested at 7% is ___?___.

7. Mrs. Chin, on reaching age 65, invested \$18,000 of her insurance policy in AA bonds paying 8½% and a common stock paying 7½%. If her yearly income from these 2 investments is \$1,470.00, how much did she invest in the bonds and how much in the stock?

8. Two investments produce a \$150 income each month. If \$1,000 more is invested at 9% than at 10%, how much was invested at each percent?

9. A man had \$12,000 to invest. If his income from a savings account earning 7½% is \$280.00 greater than his income from another investment at 8%, how much did he invest at each rate?

10. Two sums of money differing by \$2,000 were invested, the larger amount at 8½% and the smaller amount at 9½%. If the larger sum yielded an annual income that was \$110.00 more than the other, how much money was invested at each rate?

11. Mrs. John D. Rockford invested a certain amount of money at 8½%, but then invested \$600 more at 7½%. If her total income from both investments is \$4,421.00, what amount was invested at 8½% and at 7½%?

12. Mr. Robertson invested \$1,500 less at 9% than at 8%. If the income from the smaller investment is \$95.00 less than the income from the larger investment, how much did he invest at 9% and at 8%?

13. Three investments yield \$1,965.00. The amount invested at 9% is \$1,500 more than that at 8%. The amount invested at 10% is \$500 less than 3 times the amount invested at 8%. How much was invested at 8%, 9%, and 10%?

Answers to Practice Problems: Investment

1. Let $x =$ amount invested at 6%
$1{,}000 - x =$ amount invested at 7%

Equation: $.06x + .07(1{,}000 - x) = 65.00$

Move decimal point 2 places to the right

$$6x + 7{,}000 - 7x = 6{,}500$$
$$-x = 6{,}500 - 7{,}000$$
$$-x = -500$$
$$x = \$500 \text{ invested in 1st bank}$$
$$\$1{,}000 - x = \$500 \text{ invested in 2nd bank}$$

2. Let $x =$ amount invested in bonds
$12{,}000 - x =$ amount invested in common stocks

Equation: $.09x + .08(12{,}000 - x) = 1{,}040.00$

Move decimal point 2 places to the right

$$9x + 96{,}000 - 8x = 104{,}000$$
$$9x - 8x = 104{,}000 - 96{,}000$$
$$x = \$8{,}000 \text{ invested in bonds}$$
$$12{,}000 - x = \$4{,}000 \text{ invested in common stocks}$$

3. Let $x =$ amount invested at 9½%
$4{,}200 - x =$ amount invested at 9%

Equation: $.09\frac{1}{2}(x) + .09(4{,}200 - x) = \$392.00$

Move decimal point 2 places to the right

$$9\tfrac{1}{2}x + 9(4{,}200 - x) = 39{,}200$$
$$9\tfrac{1}{2}x + 37{,}800 - 9x = 39{,}200$$
$$\tfrac{1}{2}x = 39{,}200 - 37{,}800$$
$$\tfrac{1}{2}x = 1{,}400$$
$$x = \$2{,}800$$
$$4{,}200 - x = \$1{,}400$$

4. Let x = amount invested at 5%
$1{,}000 + x$ = amount invested at 6%

Equation: $.05(x) + .06(1{,}000 + x) = 1{,}710.00$

Move decimal point 2 places to the right

$$5x + 6{,}000 + 6x = 171{,}000$$
$$11x = 165{,}000$$
$$x = \$15{,}000$$
$$1{,}000 + x = \$16{,}000$$

5. Let x = amount invested at $8\frac{1}{2}$%
$14{,}000 - x$ = amount invested at $7\frac{1}{2}$%

Equation: $.08\frac{1}{2}(x) + .07\frac{1}{2}(14{,}000 - x) = 1{,}152.00$

Move decimal point 2 places to the right

$$8\tfrac{1}{2}x + 7\tfrac{1}{2}(14{,}200 - x) = 115{,}200$$
$$8\tfrac{1}{2}x + 105{,}000 - 7\tfrac{1}{2}x = 115{,}200$$
$$x = 115{,}200 - 105{,}000$$
$$x = \$10{,}200$$
$$14{,}000 - x = \$3{,}800$$

6. Let x = amount invested at 7%
$2x$ = amount invested at $6\frac{1}{2}$%

Equation: $.06\frac{1}{2}(2x) + .07(x) = 800.00$

Move decimal point 2 places to the right

$$13x + 7x = 80{,}000$$
$$20x = 80{,}000$$
$$x = \$4{,}000$$
$$2x = \$8{,}000$$

7. Let x = amount invested at $8\frac{1}{2}$% bonds
$18{,}000 - x$ = amount invested at $7\frac{1}{2}$% common stock

Equation: $.08\frac{1}{2}(x) + .07\frac{1}{2}(18{,}000 - x) = \$1{,}470.00$

Move decimal point 2 places to the right

$$8\tfrac{1}{2}(x) + 7\tfrac{1}{2}(18{,}000 - x) = 147{,}000$$
$$8\tfrac{1}{2}x + 135{,}000 - 7\tfrac{1}{2}x = 147{,}000$$
$$x = 147{,}000 - 135{,}000$$
$$x = \$12{,}000$$
$$18{,}000 - x = \$6{,}000$$

8. Let x = amount invested at 10%
$1{,}000 + x$ = amount invested at 9%
\$150 per month = 12×150 = \$1,800.00 per year

Equation: $.10(x) + .09(1{,}000 + x) = 1{,}800.00$

Move decimal point 2 places to the right

$$10(x) + 9{,}000 + 9x = 180{,}000$$
$$10x + 9{,}000 + 9x = 180{,}000$$
$$19x = 180{,}000 - 9{,}000$$
$$19x = 171{,}000$$
$$x = \$9{,}000$$
$$1{,}000 + x = \$10{,}000$$

9. Let x = amount invested at 7½%
$12{,}000 - x$ = amount invested at 8%

Equation: $.07\tfrac{1}{2}x - .08(12{,}000 - x) = 280.00$

Move decimal point 2 places to the right

$$7\tfrac{1}{2}x - 96{,}000 + 8x = 28{,}000$$
$$15\tfrac{1}{2}x = 28{,}000 + 96{,}000$$
$$15.5x = 124{,}000$$
$$x = \$8{,}000$$
$$12{,}000 - x = \$4{,}000$$

10.

1st Investment		2nd Investment		Yield
x	−	x − 2,000	=	\$110.00
8½%		9½%		

Equation: $.08\tfrac{1}{2}(x) - .09\tfrac{1}{2}(x - 2{,}000) = 110.00$

Move decimal point 2 places to the right

$$8\tfrac{1}{2}(x) - 9\tfrac{1}{2}x + 19{,}000 = 11{,}000$$
$$8\tfrac{1}{2}x - 9\tfrac{1}{2}x = 11{,}000 - 19{,}000$$
$$-x = -8{,}000$$
$$x = \$8{,}000 \text{ invested @ } 8\tfrac{1}{2}\%$$
$$x - 2{,}000 = \$6{,}000 \text{ invested @ } 9\tfrac{1}{2}\%$$

11.

1st Investment		2nd Investment		Yield
x	+	x + 600	=	\$4,421.00
8½%		7½%		

Equation: $.08\tfrac{1}{2}(x) + .07\tfrac{1}{2}(x + 600) = 4{,}421.00$

Move decimal point 2 places to the right

$$8\tfrac{1}{2}x + 7\tfrac{1}{2}x + 4{,}500 = 442{,}100$$
$$16x = 442{,}100 - 4{,}500$$
$$16x = 437{,}600$$
$$x = \$27{,}350 \text{ at } 8\tfrac{1}{2}\%$$
$$x + 600 = \$27{,}950 \text{ at } 7\tfrac{1}{2}\%$$

12.

1st Investment		2nd Investment		Yield
x	−	$x-1{,}500$	=	\$95.00
8%		9%		

Equation: $.08(x)-.09(x-1{,}500)=95.00$

Move decimal point 2 places to the right

$$8x-9(x-1{,}500)=9{,}500$$
$$8x-9x+13{,}500=9{,}500$$
$$-x+13{,}500=9{,}500$$
$$-x=9{,}500-13{,}500$$
$$x=\$4{,}000 \text{ at } 8\%$$
$$x-1{,}500=\$2{,}500 \text{ at } 9\%$$

13.

1st Investment		2nd Investment		3rd Investment		Yield
x	+	$x+1{,}500$	+	$3x-500$	=	\$1,965.00
8%		9%		10%		

Equation: $.08(x)+.09(x+1{,}500)+.10(3x-500)=1{,}965.00$

Move decimal point 2 places to the right

$$8x+9x+13{,}500+30x-5{,}000=196{,}500$$
$$47x+8{,}500=196{,}500$$
$$47x=196{,}500-8{,}500$$
$$47x=188{,}000$$
$$x=\$4{,}000 \text{ @ } 8\%$$
$$x+1{,}500=\$5{,}500 \text{ @ } 9\%$$
$$3x-500=\$11{,}500 \text{ @ } 10\%$$

8 Work Problems

Study these complex fractions and their simplification to common fractions. They will improve your ability to solve work problems.

1. $\frac{1}{2\frac{1}{2}} = 1 \div 2\frac{1}{2}$

$= 1 \div \frac{5}{2}$

$= 1 \cdot \frac{2}{5}$

$\frac{1}{2\frac{1}{2}} = \frac{2}{5}$

2. $\frac{1}{3\frac{1}{3}} = 1 \div 3\frac{1}{3}$

$= 1 \div \frac{10}{3}$

$= 1 \cdot \frac{3}{10}$

$\frac{1}{3\frac{1}{3}} = \frac{3}{10}$

3. $\frac{2x}{6\frac{2}{3}} = 2x \div 6\frac{2}{3}$

$= 2x \div \frac{20}{3}$

$= \cancel{2}x \cdot \frac{3}{\cancel{20}\ 10}$

$\frac{2x}{6\frac{2}{3}} = \frac{3x}{10}$

4. $\frac{4\frac{1}{2}}{7\frac{1}{3}} = 4\frac{1}{2} \div 7\frac{1}{3}$

$= \frac{9}{2} \div \frac{22}{3}$

$= \frac{9}{2} \cdot \frac{3}{22}$

$\frac{4\frac{1}{2}}{7\frac{1}{3}} = \frac{27}{44}$

5. $\frac{5}{11\frac{2}{3}} = 5 \div 11\frac{2}{3}$

$= 5 \div \frac{35}{3}$

$= \cancel{5} \cdot \frac{3}{\cancel{35}\ 7}$

$\frac{5}{11\frac{2}{3}} = \frac{3}{7}$

Before solving work problems, you must be able to solve equations containing fractions. In order to solve these equations, you must have a lowest common denominator (L.C.D.).

Example 1

Solve for x:

Equation: $\frac{1}{2} + \frac{1}{4} = \frac{1}{x}$

Step 1. Find the L.C.D. For this equation, L.C.D. $= 4x$.

Step 2. Multiply each numerator of each fraction by the L.C.D.

Step 3. Divide each numerator by the denominator of each fraction.

Step 4. Solve equation for the value of the letter.

Above equation: $\frac{1}{2}+\frac{1}{4}=\frac{1}{x}$

L.C.D. $= 4x$
$$\frac{{}^{2}\cancel{4}x}{\cancel{2}}+\frac{\cancel{4}x}{\cancel{4}}=\frac{4\cancel{x}}{\cancel{x}}$$
$$2x+x=4$$
$$3x=4$$
$$x=1\tfrac{1}{3}$$

Example 2

Equation:
$$\frac{15}{2x}+\frac{4}{x}=\frac{23}{4}$$

L.C.D. $= 4x$
$${}^{2}\cancel{4}\cancel{x}\cdot\frac{15}{\cancel{2}\cancel{x}}+4\cancel{x}\cdot\frac{4}{\cancel{x}}=\cancel{4}x\cdot\frac{23}{\cancel{4}}$$
$$30+16=23x$$
$$46=23x$$
$$2=x$$

Example 3

Equation:
$$\frac{3x-4}{7}=\frac{2x-4}{4}$$

L.C.D. $= 28$
$$\frac{{}^{4}\cancel{28}(3x-4)}{\cancel{7}}=\frac{{}^{7}\cancel{28}(2x-4)}{\cancel{4}}$$
$$12x-16=14x-28$$
$$12x-14x=-28+16$$
$$-2x=-12$$
$$x=6$$

Example 4

Equation:
$$\frac{3}{x+2}+\frac{5}{x-2}=\frac{28}{(x+2)(x-2)}$$

L.C.D. $= (x+2)(x-2)$
$$\cancel{(x+2)}(x-2)\frac{3}{\cancel{x+2}}+(x+2)\cancel{(x-2)}\frac{5}{\cancel{x-2}}=\frac{28\,\cancel{(x+2)}\cancel{(x-2)}}{\cancel{(x+2)}\cancel{(x-2)}}$$

$$3(x-2)+5(x+2)=28$$
$$3x-6+5x+10=28$$
$$3x+5x=28+6-10$$
$$8x=24$$
$$x=3$$

STUDY PROBLEM 1

Bill can mow the lawn in 2 hours, and his younger brother Tom can mow it in 3 hours. If both brothers mow the lawn together, how many hours will it take them?

Model:

Persons	*Hours to mow lawn*	*Part done in 1 hr*
Bill	2	1/2
Tom	3	1/3
Together	x	$1/x$

Equation: $$\frac{1}{2}+\frac{1}{3}=\frac{1}{x}$$

L.C.D. $=6x$

$$3x+2x=6$$
$$5x=6$$
$x=1\frac{1}{5}$ hr or 1 hr and 12 min to finish the job together

STUDY PROBLEM 2

One pump can fill a swimming pool in 8 hours and another pump can fill it in 10 hours. If both pumps are opened at the same time, how many hours will it take to fill the pool?

Equipment	*Hours to fill pool*	*Part filled in 1 hr*
1st pump	8	1/8
2nd pump	10	1/10
Together	x	$1/x$

Equation: $$\frac{1}{8}+\frac{1}{10}=\frac{1}{x}$$

L.C.D. $=40x$

$$5x+4x=40$$
$$9x=40$$
$x=4\frac{4}{9}$ hr

STUDY PROBLEM 3

One pump can fill a reservoir in 60 hours. Another pump can fill the same reservoir in 80 hours. A third pump can empty the reservoir in 90 hours. If all three pumps are operating at the same time, how long will it take to fill the reservoir?

Equipment	*Hours to fill reservoir*	*Part filled in 1 hr*
1st pump	60	1/60
2nd pump	80	1/80
3rd pump	90	1/90
Together	x	$1/x$

Equation: $$\frac{1}{60}+\frac{1}{80}-\frac{1}{90}=\frac{1}{x}$$

L.C.D. $= 720x$

$$12x+9x-8x=720$$
$$13x=720$$
$$x=55\tfrac{5}{13}\text{ hr}$$

STUDY PROBLEM 4

One printing press can print 5,000 advertising cards in 12 seconds. Another printing press can print the same number of cards in 7½ seconds. If both presses are used together to print the 5,000 cards, how many seconds will it take them?

Equipment	*Seconds to print job*	*Part done in 1 sec*
1st press	12	1/12
2nd press	7½	1/7½
Together	x	$1/x$

Equation: $\frac{1}{12}+\frac{1}{7\frac{1}{2}}=\frac{1}{x}$

L.C.D. $= 60x$

Multiply each numerator by the L.C.D.

Divide each numerator by each denominator.

Solve equation.

$$\frac{1}{12} + \frac{2}{15} = \frac{1}{x}$$

$${}^{5}\cancel{60}x\frac{(1)}{\cancel{12}} + {}^{4}\cancel{60}x\frac{(2)}{\cancel{15}} = 60\cancel{x}\frac{(1)}{\cancel{x}}$$

$$5x + 8x = 60$$

$$13x = 60$$

$$x = 4\tfrac{8}{13} \text{ sec}$$

STUDY PROBLEM 5

Willard can paint a living room in 6 hours. His assistant can paint the same room in 8 hours. If Willard paints for 1 hour and then leaves, letting his assistant take over, how long will it take the assistant to complete the job?

Note: Problems that involve 1 or more workers who perform at various rates of speed and are then replaced by other workers can be solved with the following format.

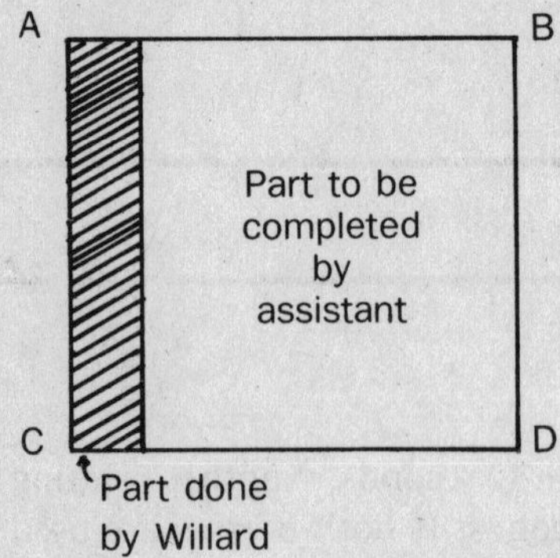

ABCD is a square that represents the whole job. The shaded area represents 1/6 of the job. The blank area, 5/6 of the job.

Persons	*Hours to complete job*	*Part done in 1 hr*
Willard	6	1/6
Assistant	8	1/8
Together	x	$1/x$
Part to be completed ------------------ 5/6 of job		

Since Willard left the job, he is no longer in the equation.

Equation: $\frac{1}{8} = \frac{1}{x} \cdot \frac{5}{6}$ Part to be completed by assistant

L.C.D. = $24x$ $\quad {}^{3}\cancel{24}x\left(\frac{1}{\cancel{8}}\right) = {}^{4}\cancel{24}\cancel{x}\left(\frac{5}{\cancel{6}\cancel{x}}\right)$

$$3x = 20$$

$x = 6\tfrac{2}{3}$ hr to complete the job

STUDY PROBLEM 6

In the same problem as above, if Willard worked 2 hours and then left the job,

a. What fraction of the job did he paint?

b. What fraction of the job has to be painted by the assistant?

c. The equation is $\frac{1}{8} = \frac{1}{x} \cdot \frac{?}{?} \rightarrow$ What fraction?

d. L.C.D. = ?

e. $x = ?$

See answers on page 58.

STUDY PROBLEM 7

One computer can do a printout for a job in 7 hours. Another computer can do the same printout in 8 hours. If the first computer is used for 2 hours and then the second computer is used, how long will it take the second computer to complete the job?

Equipment	*Hours to complete job*	*Part done in 1 hr*
1st computer	7	1/7
2nd computer	8	1/8
Together	x	$1/x$
Part to be completed-------------------------		5/7 of job

Equation: $\frac{1}{8} = \frac{1}{x} \cdot \frac{5}{7} \rightarrow$ Part to be completed

L.C.D. = $56x$

$7x = 40$

$x = 5\frac{5}{7}$ hr for 2nd computer

STUDY PROBLEM 8

John can paint a fence alone in 30 minutes. If John and Bill together can complete the job in 18 minutes, how long will it take Bill to complete the job alone?

Persons	*Minutes to complete job*	*Part done in 1 min*
Bill	x	$1/x$
John	30	1/30
Together	18	1/18

Equation: $\frac{1}{30}+\frac{1}{x}=\frac{1}{18}$

L.C.D. $=90x$

$$3x+90=5x$$
$$90=5x-3x$$
$$90=2x$$
$$45=x \text{ min}$$

STUDY PROBLEM 9

Bess can type an attorney's brief in 6 hours. Betty can type the same brief in 8 hours. If Bess works 1 hour and stops, and then Betty continues to complete the job, how long will it take Betty to complete the job alone?

Persons	*Hours to complete job*	*Part done in 1 hr*
Bess	6	1/6
Betty	8	1/8
Together	x	$1/x$
Part to be completed------------------5/6 of job		

Equation: $\frac{1}{8}=\frac{1}{x}\cdot\frac{5}{6}\rightarrow$ Part to be completed

$$\frac{1}{8}=\frac{5}{6x}$$

L.C.D. $=24x$

$$3x=20$$
$x=6⅔$ hr for Betty to complete the job

Answers to Study Problem 6

a. 2/6 or 1/3

b. 4/6 or 2/3

c. $1/8=\frac{1}{x}\cdot\frac{2}{3}\rightarrow\frac{2}{3x}$

d. L.C.D. $=24x$

e. $3x=16$

$x=5⅓$ hr or 5 hr and 20 min

Practice Problems: Work

1. It takes Mr. Rose 10 hours to complete a printing job. His helper John can complete the same job in 15 hours. If Mr. Rose and John work together, how long will it take them to complete the job?

2. It takes one machine 25 minutes to pick a bale of cotton and another machine 30 minutes. If both machines are used, how long will it take them to pick a bale of cotton?

3. A tank can be filled by one pump in 50 minutes and by another pump in 60 minutes. A third pump can drain the tank in 75 minutes. If all 3 pumps go into operation, how long will it take to fill the tank?

4. A gardener can care for the Green's property in 5 hours. If his helper assists him, they can complete the job in 4 hours. How long will it take the helper to do the job alone?

5. Walter, Bill, and Tom want to decorate the multi-purpose room in school for a dance. It would take Walter and Bill 4 hours each to do the job and Tom alone could do the job in 3 hours. If all work together, how long will it take them to decorate the room?

6. Nancy can do a typing job in 8 hours. When Carole helps her, they can do the job together in 5 hours. How many hours would it take Carole to do the job alone?

7. Mr. Sharrone can paper a room in 3 hours. His assistant needs 5 hours. If Mr. Sharrone works 1 hour and then his assistant completes the job alone, how long will it take the assistant to finish the job?

8. Mr. Rodriguez, a college instructor, can grade his class papers in 3 hours while it takes his assistant 4½ hours. If Mr. Rodriguez graded the papers for 1 hour and then left the job for his assistant to complete, how long will it take his assistant to finish grading papers?

9. It takes 8 hours for one crew to clean an office building. It takes a second crew 10 hours to clean the same building. If the first crew works 2 hours and leaves and the second crew takes over to complete the job, how long will it take the second crew to finish cleaning the building?

10. Mr. Ito's assistant takes twice as long to complete a computer task as Mr. Ito. If it takes both experts 6 hours to complete the task, how long will it take each of them to do the job alone?

11. Mr. Vella can build a brick wall in 4 days. His apprentice can build the same wall in 6 days. After working alone for 3 days, Mr. Vella became ill and left the job for his apprentice to complete. How many days did it take the apprentice to finish the wall?

Answers to Practice Problems: Work

1.

Persons	*Hours to complete job*	*Part done in 1 hr*
Mr. Rose	10	1/10
John	15	1/15
Together	x	$1/x$

Equation: $$\frac{1}{10}+\frac{1}{15}=\frac{1}{x}$$

L.C.D. $= 30x$

$$3x+2x=30$$
$$5x=30$$
$$x=6 \text{ hr to complete job together}$$

2.

Equipment	*Minutes to pick bale*	*Part done in 1 hr*
1st machine	25	1/25
2nd machine	30	1/30
Together	x	$1/x$

Equation: $$\frac{1}{25}+\frac{1}{30}=\frac{1}{x}$$

L.C.D. $= 150x$

$$6x+5x=150$$
$$11x=150$$
$$x=13\tfrac{7}{11} \text{ min to pick a bale}$$

3.

Equipment	*Minutes to fill tank*	*Part filled in 1 min*
1st pump	50	1/50
2nd pump	60	1/60
3rd pump	75 drain (−)	−1/75
Together	x	$1/x$

Equation: $$\frac{1}{50}+\frac{1}{60}-\frac{1}{75}=\frac{1}{x}$$

L.C.D. $= 300x$

$$6x+5x-4x=300$$
$$7x=300$$
$$x=42\tfrac{6}{7} \text{ min to fill tank}$$

4.

Persons	*Hours to complete job*	*Part done in 1 hr*
Gardener	5	1/5
Helper	x	$1/x$
Together	4	1/4

Equation: $$\frac{1}{5}+\frac{1}{x}=\frac{1}{4}$$

L.C.D. $= 20x$

$$4x+20=5x$$
$$20=5x-4x$$
$20 = x$ hr for helper to complete the job alone

5.

Persons	*Hours to complete job*	*Part done in 1 hr*
Walter	4	1/4
Bill	4	1/4
Tom	3	1/3
Together	x	$1/x$

Equation: $$\frac{1}{4}+\frac{1}{4}+\frac{1}{3}=\frac{1}{x}$$

L.C.D. $= 12x$

$$3x+3x+4x=12$$
$$10x=12$$
$$x=1\tfrac{2}{10}$$
$x = 1\tfrac{1}{5}$ hr if all work together

6.

Persons	*Hours to complete job*	*Part done in 1 hr*
Nancy	8	1/8
Carole	x	$1/x$
Together	5	1/5

Equation: $$\frac{1}{8}+\frac{1}{x}=\frac{1}{5}$$

L.C.D. $= 40x$

$$5x+40=8x$$
$$40=8x-5x$$
$$40=3x$$
$13\tfrac{1}{3} = x$ hr for Carole to do the job alone

7.

Persons	*Hours to complete job*	*Part done in 1 hr*
Sharrone	3	1/3
Assistant	5	1/5
Together	x	$1/x$
Part to be completed --------------------- 2/3 of job		

Equation: $\frac{1}{5} = \frac{1}{x} \cdot \frac{2}{3}$

L.C.D. $= 15x$

$$\frac{1}{5} = \frac{2}{3x}$$

$3x = 10$

$x = 3\frac{1}{3}$ hr for assistant to complete job

8.

Persons	*Hours to complete job*	*Part done in 1 hr*
Mr. Rodriguez	3	1/3
Assistant	$4\frac{1}{2}$	2/9
Together	x	$1/x$
Part to be completed -------------------------- 2/3 of job		

Equation: $\frac{2}{9} = \frac{1}{x} \cdot \frac{2}{3}$

L.C.D. $= 9x$

$$\frac{2}{9} = \frac{2}{3x}$$

$2x = 6$

$x = 3$ hr for assistant to finish grading papers

9.

Persons	*Hours to complete job*	*Part done in 1 hr*
1st crew	8	1/8
2nd crew	10	1/10
Together	x	$1/x$
Part to be completed ----------------- 3/4 of job		

Equation: $\frac{1}{10} = \frac{1}{x} \cdot \frac{3}{4}$

$$\frac{1}{10} = \frac{3}{4x}$$

L.C.D. $= 20x$

$2x = 15$

$x = 7\frac{1}{2}$ hr for 2nd crew to finish

10.

Persons	*Hours to complete job*	*Part done in 1 hr*
Mr. Ito	x	$1/x$
Assistant	$2x$	$1/2x$
Together	6	$1/6$
Part to be completed ----------------------		whole job

Equation: $\frac{1}{x}+\frac{1}{2x}=\frac{1}{6}$

L.C.D. $=6x$

$$6+3=x$$

$$x=9 \text{ hr for Mr. Ito}$$

$$2x=18 \text{ hr for assistant}$$

11.

Persons	*Days to do job*	*Part done in 1 day*
Mr. Vella	4	$1/4$
Apprentice	6	$1/6$
Together	x	$1/x$
Part to be completed --------------		1/4 of job

Equation: $\frac{1}{6}=\frac{1}{x}\cdot\frac{1}{4}$

L.C.D. $=12x$

$$^{2}\cancel{12}x\left(\frac{1}{\cancel{6}}\right)=\left(\frac{1}{\cancel{4}\cancel{x}}\right){}^{3}\cancel{12}\cancel{x}$$

$$2x=3$$

$x=1\frac{1}{2}$ days for apprentice to complete the job

9 Attendance Problems

To solve problems that involve attendance at events, you can use the same model used in solving investment problems.

Model:

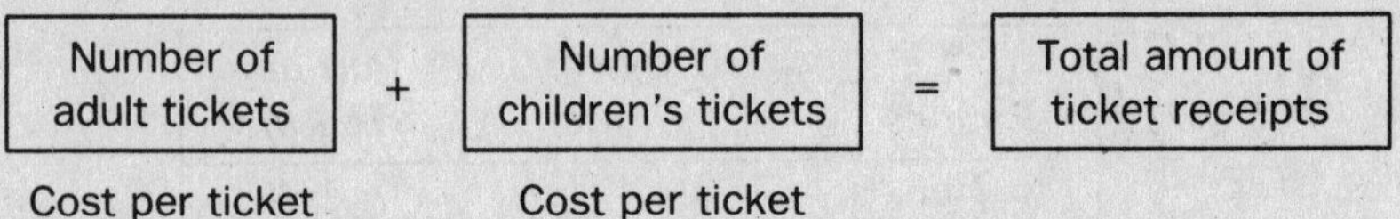

STUDY PROBLEM 1

The admission at a concert was \$7.50 for adults and \$3.00 for children. If the total gate receipts were \$4,050 and there were 600 people in attendance, how many adult tickets and how many children's tickets were sold?

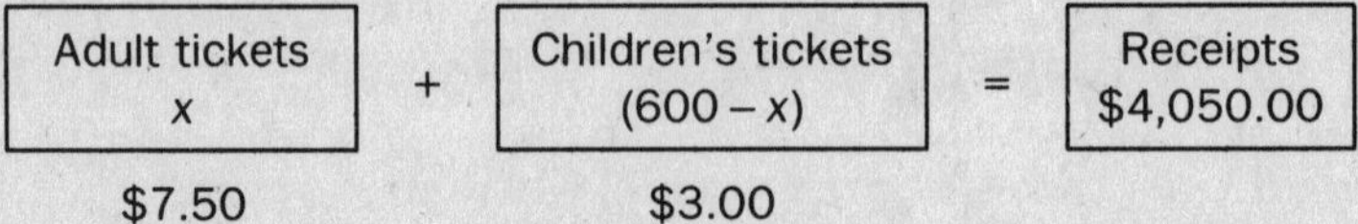

a. Move decimal point 2 places to the right.
b. Multiply cost of each ticket by the number of tickets.

Equation: $7.50(x)+3.00(600-x)=4,050.00$

$$750x+300(600-x)=405,000$$
$$750x+180,000-300x=405,000$$
$$450x=405,000-180,000$$
$$450x=225,000$$
$$x=500$$
$$600-x=100$$

Practice Problems: Attendance

1. The admission at a local football game was \$3.50 for adults and \$1.25 for students and children. The receipts were \$4,575.00 for 1,500 admissions. How many of each kind of ticket were sold?

2. On opening night, 1,500 tickets were sold for a college play. Adult tickets were sold at \$5.25 each and student tickets at \$2.25 each. If the total receipts collected were \$4,875.00, how many of each kind of ticket were sold?

3. A civic auditorium has 1,300 seats. At a holiday community function, all seats were sold. Adult tickets cost \$3.50 each and children's tickets cost \$2.00 each. If the total amount collected was \$3,800.00, how many of each kind of ticket were sold?

4. At a recent ballet school performance, the number of young adult tickets sold was 50 less than twice the number of adult tickets sold, and the number of children's tickets sold was the same as the number of adult tickets. If adult tickets cost \$2.50 each, young adult tickets \$1.50 each, and children's tickets \$1.00 each, how many of each kind of ticket were sold if the cash receipts amounted to \$705.00?

5. At a university football game, 5,000 more regular-seat tickets were sold at \$7.50 each than box office seats at \$12.50 each. If the receipts collected amounted to \$637,000, how many tickets were sold at \$7.50 each and how many were sold at \$12.50 each?

See answers that follow.

Answers to Practice Problems: Attendance

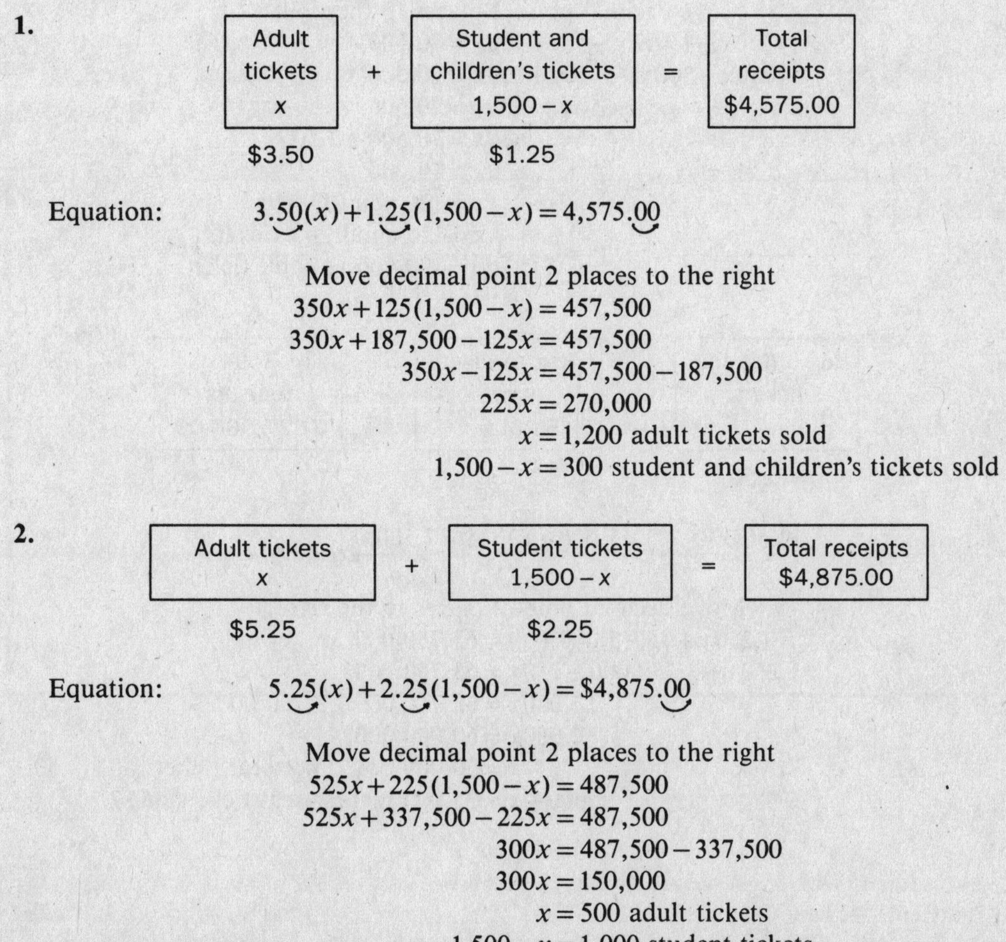

1. Equation: $3.50(x)+1.25(1{,}500-x)=4{,}575.00$

Move decimal point 2 places to the right

$$350x+125(1{,}500-x)=457{,}500$$
$$350x+187{,}500-125x=457{,}500$$
$$350x-125x=457{,}500-187{,}500$$
$$225x=270{,}000$$
$$x=1{,}200 \text{ adult tickets sold}$$
$$1{,}500-x=300 \text{ student and children's tickets sold}$$

2. Equation: $5.25(x)+2.25(1{,}500-x)=\$4{,}875.00$

Move decimal point 2 places to the right

$$525x+225(1{,}500-x)=487{,}500$$
$$525x+337{,}500-225x=487{,}500$$
$$300x=487{,}500-337{,}500$$
$$300x=150{,}000$$
$$x=500 \text{ adult tickets}$$
$$1{,}500-x=1{,}000 \text{ student tickets}$$

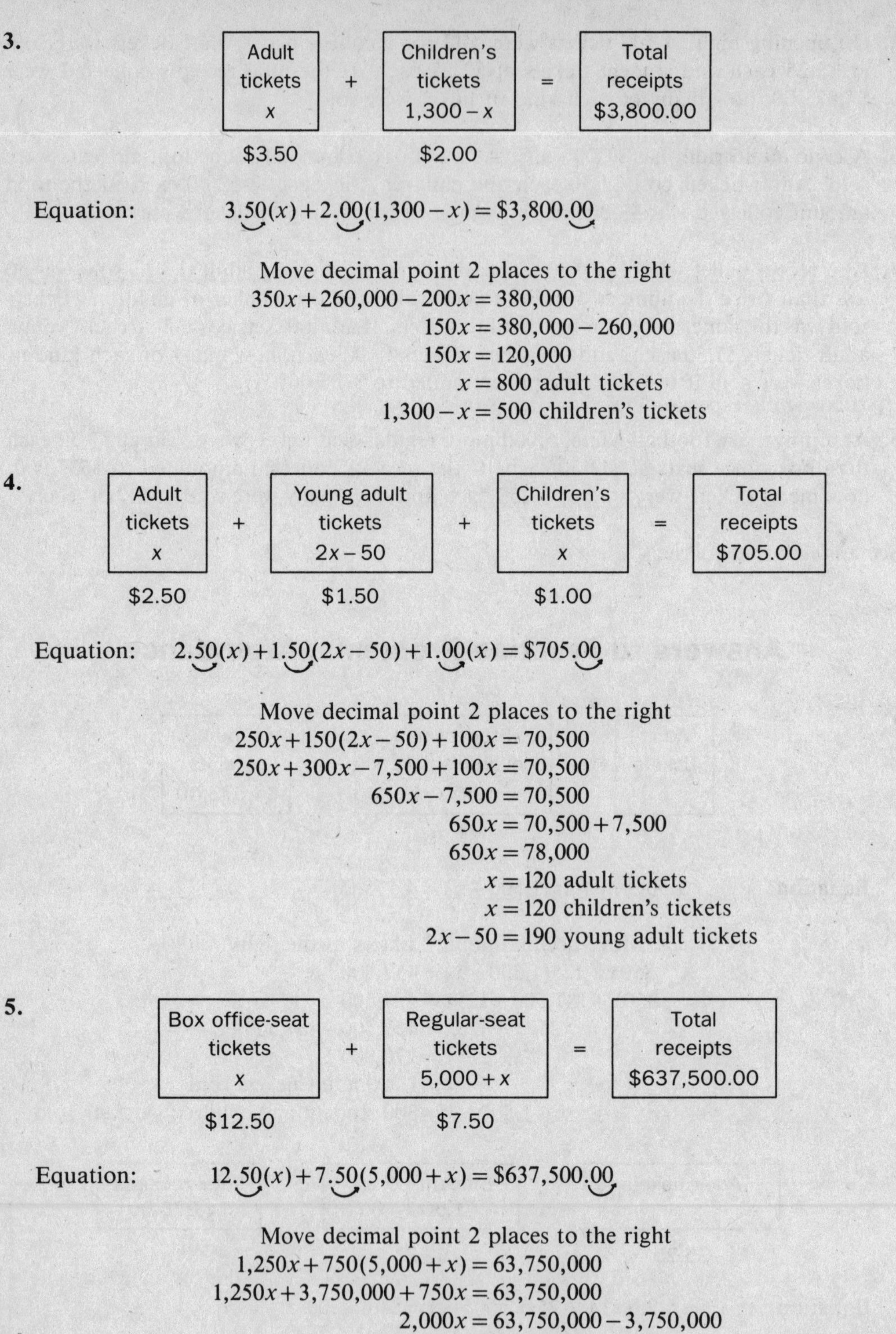

3.

Equation: $3.50(x)+2.00(1{,}300-x)=\$3{,}800.00$

Move decimal point 2 places to the right

$$350x+260{,}000-200x=380{,}000$$
$$150x=380{,}000-260{,}000$$
$$150x=120{,}000$$
$$x=800 \text{ adult tickets}$$
$$1{,}300-x=500 \text{ children's tickets}$$

4.

Equation: $2.50(x)+1.50(2x-50)+1.00(x)=\$705.00$

Move decimal point 2 places to the right

$$250x+150(2x-50)+100x=70{,}500$$
$$250x+300x-7{,}500+100x=70{,}500$$
$$650x-7{,}500=70{,}500$$
$$650x=70{,}500+7{,}500$$
$$650x=78{,}000$$
$$x=120 \text{ adult tickets}$$
$$x=120 \text{ children's tickets}$$
$$2x-50=190 \text{ young adult tickets}$$

5.

Equation: $12.50(x)+7.50(5{,}000+x)=\$637{,}500.00$

Move decimal point 2 places to the right

$$1{,}250x+750(5{,}000+x)=63{,}750{,}000$$
$$1{,}250x+3{,}750{,}000+750x=63{,}750{,}000$$
$$2{,}000x=63{,}750{,}000-3{,}750{,}000$$
$$2{,}000x=60{,}000{,}000$$
$$x=30{,}000 \text{ box office-seat tickets at \$12.50}$$
$$5{,}000+x=35{,}000 \text{ regular-seat tickets at \$7.50}$$

10 Lever Problems

A diagram is especially helpful when working with lever problems.

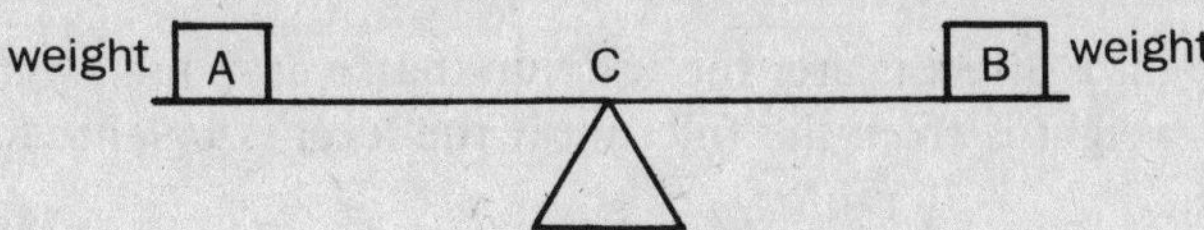

A lever is a stiff board or bar which is balanced upon a support called a fulcrum. AC and CB are called lever arms, and C is the fulcrum. If weight A and weight B are equal, AC will balance CB if C, the fulcrum, is in the middle. If you played on a teeterboard or seesaw when you were a child, you know that two people will balance only when the heavier of the two sits closer to the fulcrum.

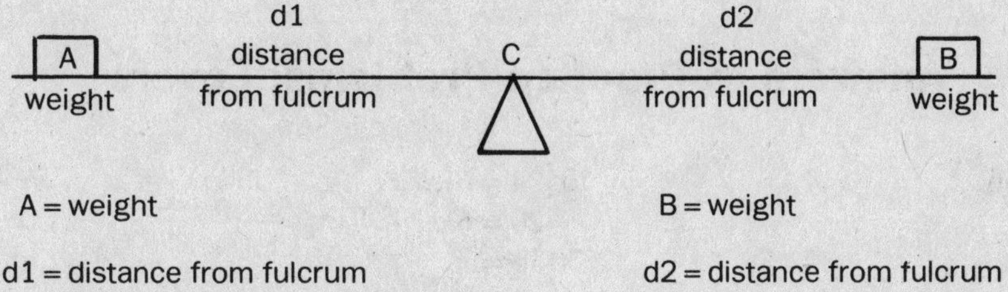

A = weight

B = weight

d1 = distance from fulcrum

d2 = distance from fulcrum

FORMULA: $W(A) \cdot d1 = d2 \cdot W(B)$

STUDY PROBLEM 1

A 40-pound weight placed 2 feet from the fulcrum of a lever balances an unknown weight placed 4 feet from the fulcrum. Find the unknown weight.

Diagram:

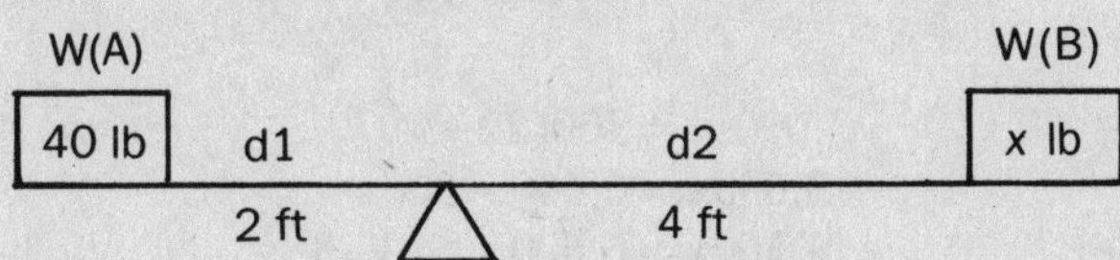

Equation:

$$40 \cdot 2 = 4 \cdot x$$
$$80 = 4x$$
$$x = 20 \text{ lb}$$

Practice Problems: Levers

1. A weight of 105 pounds is 4 feet from the fulcrum of a lever, and an unknown weight is 6 feet from the fulcrum. What is the unknown weight if the lever is balanced?

2. A 90-pound weight is placed on a lever at 7 feet from the fulcrum. At what distance from the fulcrum must a 63-pound weight be placed to make the lever balance?

3. A 100-pound weight is 3 feet nearer the fulcrum than a 70-pound weight. Find the distances that each weight is from the fulcrum if the lever is balanced.

4. A 3,000-pound weight rests on one end of a lever and a 300-pound weight rests on the other end. If the lever is 22 feet long, where must the fulcrum be placed to make the lever balance?

5. A heavy boulder had to be removed from a plot of land so that the builder could begin excavating for the foundation of a new home. The builder decided to use the lever principle to move the boulder. If the boulder is 1 foot from the fulcrum and a 300-pound weight is 3 feet from the fulcrum, what is the weight of the boulder? (Draw a diagram.)

Answers to Practice Problems: Levers

1. Equation:

$$105 \cdot 4 = 6 \cdot x$$
$$420 = 6x$$
$$70 \text{ lb} = x$$

2. Equation:

$$90 \cdot 7 = x \cdot 63$$
$$630 = 63x$$
$$10 \text{ ft} = x$$

3. Equation:

$$100(x-3) = x \cdot 70$$
$$100x - 300 = 70x$$
$$100x - 70x = 300$$
$$30x = 300$$
$$x = 10 \text{ ft for the 70-lb weight}$$
$$x - 3 = 7 \text{ ft for the 100-lb weight}$$

4. Equation:

$$3{,}000 \cdot x = 300(22 - x)$$
$$3{,}000x = 6{,}600 - 300x$$
$$3{,}300x = 6{,}600$$
$$x = 2 \text{ ft from the 3,000-lb weight}$$

5.

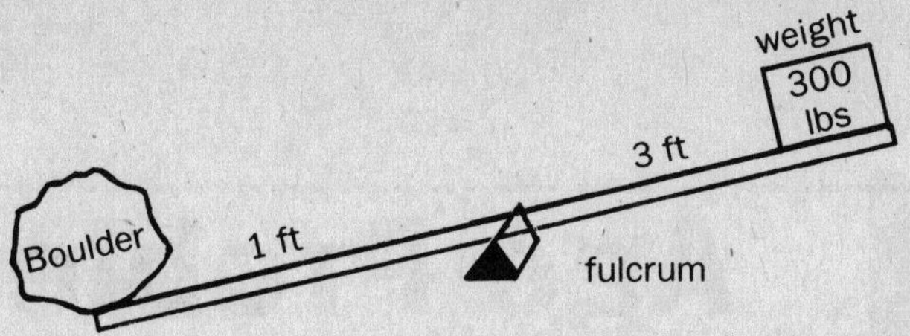

$$W \cdot 1 = 3 \cdot (300)$$
$W1 = 900$-lb weight of boulder

11 Age Problems

It's important to understand the translation of English into algebraic symbols. For example:

I. a. Four years from now = +4

b. Four years ago = −4

c. Sherrie is 3 times as old as her sister Susie. Five years from now, Sherrie will be twice as old as Susie. How old is each now?

Fill in the question marks:

Model:

Persons	*Age now*	*Age 5 years from now*
Sherrie	?	?
Susie	x	?

Equation: $? = 2(?)$

II. If Bill's age now is x years, represent the following algebraically:

a. Bill's age 4 years from now

b. Bill's age 6 years ago

c. ⅔ of Bill's age 1 year ago

d. ¾ of Bill's age 2 years from now

III. The sum of Anne's age and her mother's age is 42 years. If x is Anne's age, her mother's age is $42-x$ years.

Answers to Age Problems I and II

I.

Persons	*Age now*	*Age 5 years from now*
Sherrie	$3x$	$3x+5$
Susie	x	$x+5$

Equation:

$$3x+5=2(x+5)$$
$$3x+5=2x+10$$
$$x=5 \text{ (Susie's age)}$$
$$3x=15 \text{ (Sherrie's age)}$$

II. a. $x+4$
b. $x-6$
c. $\frac{2}{3}(x-1)$
d. $\frac{3}{4}(x+2)$

STUDY PROBLEM 1

Juan is 5 years older than his brother Carlos. Five years ago, Juan was twice as old as Carlos. How old is each now?

Persons	*Age now*	*Age 5 years ago*
Juan	$x+5$	$x+5-5$
Carlos	x	$x-5$

Equation:

$$x+5-5=2(x-5)$$
$$x=2(x-5)$$
$$x=2x-10$$
$$10=2x-x$$
$$10=x \text{ (Carlos's age now)}$$
$$15=x+5 \text{ (Juan's age now)}$$

STUDY PROBLEM 2

The sum of John's age and his brother Jerry's age is 42. Eight years ago John's age was 2 years more than Jerry's. How old is each now?

Persons	*Age now*	*Age 8 years ago*
John	$42-x$	$42-x-8$
Jerry	x	$x-8$

Equation:

$$42-x-8=x-8+2$$
$$34-x=x-6$$
$$-x-x=-6-34$$
$$-2x=-40$$
$$x=20 \text{ (Jerry's age)}$$
$$42-x=22 \text{ (John's age)}$$

STUDY PROBLEM 3

Kathryn is 8 years older than her sister Rosemary. Eight years ago Kathryn was 3 times as old as Rosemary. How old is each now?

Model:

Persons	*Age now*	*Age 8 years ago*
Kathryn	$x+8$	$x+8-8$
Rosemary	x	$x-8$

Equation:

$$\begin{aligned} x &= 3(x-8) \\ x &= 3x-24 \\ x-3x &= -24 \\ -2x &= -24 \\ x &= 12 \text{ (Rosemary's age now)} \\ x+8 &= 20 \text{ (Kathryn's age now)} \end{aligned}$$

STUDY PROBLEM 4

Mr. Wong is 4 times as old as his daughter Sylvia. In 10 years Mr. Wong will be 10 times as old as his daugher was 5 years ago. How old is each now?

Model:

Persons	*Age now*	*Age 10 years from now*	*Age 5 years ago*
Mr. Wong	$4x$	$4x+10$	?
Sylvia	x	?	$x-5$

Equation:

$$\begin{aligned} 4x+10 &= 10(x-5) \\ 4x+10 &= 10x-50 \\ 4x-10x &= -50-10 \\ -6x &= -60 \\ x &= 10 \text{ (Sylvia's age now)} \\ 4x &= 40 \text{ (Mr. Wong's age now)} \end{aligned}$$

Practice Problems: Age

1. Mr. Kim is 25 years older than his daughter. In 10 years Mr. Kim will be twice as old as his daughter will be. How old is each now?

2. Mr. Polaski is 5 times as old as his son David. Five years ago the father was 15 times as old as his son. What are their ages now?

3. The sum of Michael's age and his pal Ronnie's age is 41. Six times Michael's age 10 years ago was equal to 2 more than twice Ronnie's age now. How old is each now?

4. Arnold is 10 years older than his brother Richard. In 10 years Richard will be ¾ as old as Arnold. How old is each now?

5. Carole's age 8 years ago was 5 times Carl's age. Five years ago Carole's age was 3 times Carl's age. How old is each now?

6. Mrs. Tracey is now 3 times as old as her daughter Juliana. Five years from now, Mrs. Tracey will be 7 times her daughter's age 5 years ago. What are the mother's and daughter's ages now?

7. If Tina's age 1 year from now plus her age 2 years ago plus three times her age 5 years ago is equal to 69 years, how old is she now?

Check your answers with the solutions that follow.

Answers to Practice Problems: Age

1.

Persons	*Age now*	*Age in 10 years*
Mr. Kim	$x+25$	$x+35$
Daughter	x	$x+10$

Equation:

$$\begin{aligned} x+35 &= 2(x+10) \\ x+35 &= 2x+20 \\ -2x+x &= -35+20 \\ -x &= -15 \\ x &= 15 \text{ (daughter's age)} \\ x+25 &= 40 \text{ (Mr. Kim's age)} \end{aligned}$$

2.

Persons	*Age now*	*Age 5 years ago*
Mr. Polaski	$5x$	$5x-5$
David	x	$x-5$

Equation:

$$\begin{aligned} 5x-5 &= 15(x-5) \\ 5x-5 &= 15x-75 \\ 5x-15x &= -70 \\ -10x &= -70 \\ x &= 7 \text{ (David's age)} \\ 5x &= 35 \text{ (Mr. Polaski's age)} \end{aligned}$$

3.

Persons	*Age now*	*Age 10 years ago*
Michael	x	$x-10$
Ronnie	$41-x$	$41-x-10$

Equation:

$$\begin{aligned} 6(x-10) &= 2(41-x)+2 \\ 6x-60 &= 82-2x+2 \\ 6x+2x &= 84+60 \end{aligned}$$

$$8x = 144$$
$$x = 18 \text{ (Michael's age now)}$$
$$41 - x = 23 \text{ (Ronnie's age now)}$$

4.

Persons	*Age now*	*Age in 10 years*
Arnold	$x+10$	$x+20$
Richard	x	$x+10$

Equation: $x + 10 = \frac{3}{4}(x + 20)$

L.C.D. = 4

$$4x + 40 = 3x + 60$$
$$4x - 3x = 60 - 40$$
$$x = 20 \text{ (Richard's age now)}$$
$$x + 10 = 30 \text{ (Arnold's age now)}$$

5.

Persons	*Age now*	*Age 8 years ago*	*Age 5 years ago*
Carole	$5x+8$	$5x$	$5x+3$
Carl	$x+8$	x	$x+3$

Equation: $5x + 3 = 3(x + 3)$

$$5x + 3 = 3x + 9$$
$$5x - 3x = 9 - 3$$
$$2x = 6$$
$$x = 3 \text{ (Carl's age 8 years ago)}$$
$$x + 8 = 11 \text{ (Carl's age now)}$$
$$5x + 8 = 23 \text{ (Carole's age now)}$$

6.

Persons	*Age now*	*Age 5 yr from now*	*Age 5 yr ago*
Mrs. Tracey	$3x$	$3x+5$	?
Juliana	x	?	$x-5$

Equation: $3x + 5 = 7(x - 5)$

$$3x + 5 = 7x - 35$$
$$3x - 7x = -35 - 5$$
$$-4x = -40$$
$$x = 10 \text{ (Juliana's age now)}$$
$$3x = 30 \text{ (Mrs. Tracey's age now)}$$

7.

Tina's age now	*1 yr from now*	*2 yr ago*	*5 yr ago*
x	$x+1$	$x-2$	$x-5$

Equation: $(x + 1) + (x - 2) + 3(x - 5) = 69$

$$x + 1 + x - 2 + 3x - 15 = 69$$
$$5x - 16 = 69$$
$$5x = 69 + 16$$
$$5x = 85$$
$$x = 17 \text{ (Tina's age now)}$$

12 Ratio Problems

Ratio is the relationship in quantity, amount or size between two or more things. We can compare numbers rather easily. For example, if John has 10 dollars and Bill has 5 dollars, then we can say that John has 5 dollars more than Bill or we can say that he has twice as much as Bill. We can express this comparison or ratio in a variety of ways:

$$\frac{2}{1}$$

$$2:1$$

$$2 \text{ to } 1$$

$$2 \div 1$$

When we compare two rational numbers by division, we have a ratio. (A rational number is the quotient of any two integers* divided by any number except zero.)

Examples

a. $\frac{2}{5}$ d. $\frac{.03}{6.25}$ g. $\frac{-18 \text{ gal}}{+2.5 \text{ gal}}$

b. $\frac{13}{5}$ e. $\frac{-12.50}{-6.25}$ h. $\frac{-12\frac{1}{2} \text{ yd}}{-.6 \text{ yd}}$

c. $\frac{126}{17}$ f. $\frac{+412 \text{ cm}}{-4 \text{ cm}}$ i. $\frac{+16\frac{2}{3}}{-6\frac{2}{3}}$

All ratios must be reduced to their lowest terms and have the same unit of measure.

Examples

a. $\frac{12}{18} = \frac{2}{3}$

b. $\frac{25}{60} = \frac{5}{12}$

c. $\frac{-4 \text{ ft}}{+30 \text{ in.}} = \frac{-48 \text{ in.}}{30 \text{ in.}} = \frac{-8}{+5}$

d. $\frac{-2 \text{ dollars}}{-60 \text{ cents}} = \frac{200 \text{ cents}}{60 \text{ cents}} = \frac{+10}{+3}$

e. $\frac{2 \text{ yd}}{\frac{1}{2} \text{ yd}} = 2 \div \frac{1}{2} = 2 \cdot \frac{2}{1} = \frac{4}{1}$

f. $\frac{-150 \text{ cm}}{-2 \text{ meters}} = \frac{150 \text{ cm}}{200 \text{ cm}} = \frac{+3}{+4}$

*An integer is any number that is a member of a set of positive and negative whole numbers including zero.

g. $\frac{4\frac{1}{2}\text{ gal}}{4\text{ qt}} = \frac{18\text{ qt}}{4\text{ qt}} = \frac{9}{2}$

h. $\frac{25\text{ min}}{1\text{ hr}} = \frac{25\text{ min}}{60\text{ min}} = \frac{5}{12}$

i. $\frac{3x\text{ ft}}{5x\text{ ft}} = \frac{3}{5}$

Practice Problems: Ratios

Write the following ratios and reduce to lowest terms.

1. $\frac{32}{18}$

2. 27 : 81

3. 330 to 99

4. 16 ft to 4 yds

5. 38 cm : 57 cm

6. 300 m to 1 km

7. 108 in. : 3 ft

8. 12½ gal to 2 qt

9. 50% to 25%

10. 33⅓% to 66⅔%

11. 3 gal and 2 qt to 5 gal

12. 30 days : 4 weeks

13. Write another ratio which would have the same ratio as 10 : 13.

14. Golf balls formerly sold at $1.00 each. Now on sale, they sell for $9.00 a dozen. What is the ratio of the former price to the sale price?

15. Express as a ratio the number of 22-cent stamps you can buy if you have (C) cents.

16.

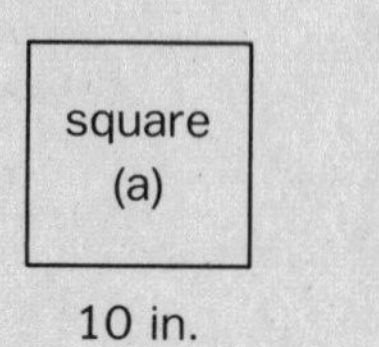

square
(b)
18 in.

What is the ratio of the perimeter of square (a) to the perimter of square (b)?

17.

rectangle
(a)
6 ft
10 ft

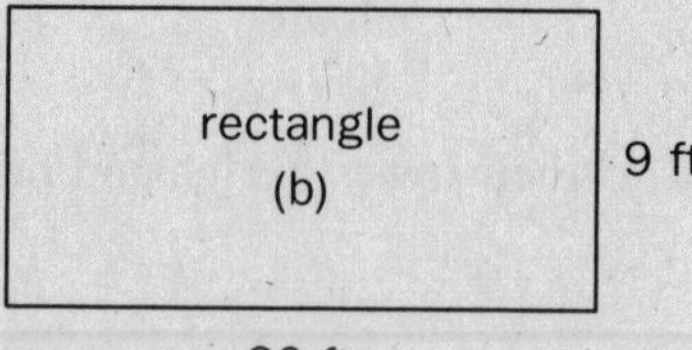

What is the ratio of the perimeter of rectangle (a) to the perimeter of rectangle (b)?

18.

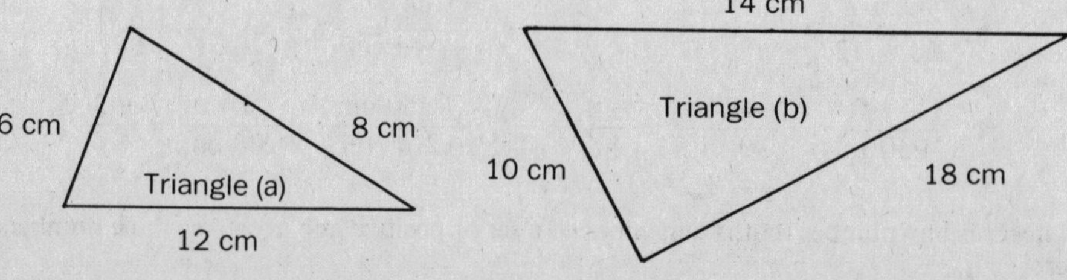

What is the ratio of the perimeter of triangle (b) to the perimeter of triangle (a)?

19. 6½ in. 6¼ in. 5 in.

Hexagon (a)

8¼ in. 7 in. 8 in.

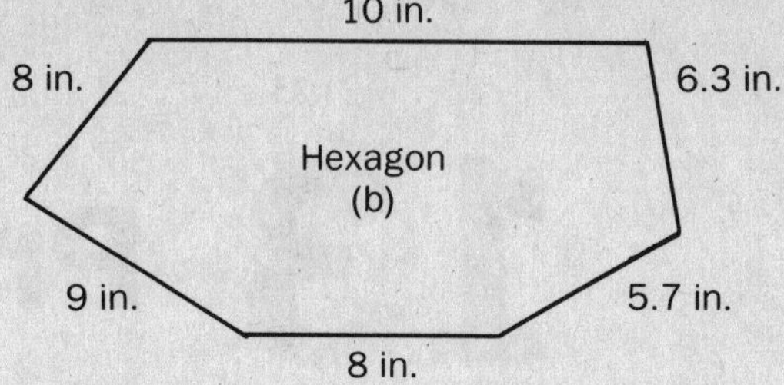

What is the ratio of the perimeter of hexagon (a) to the perimeter of hexagon (b)?

20. The budget of a small city is $15,000,000 for the year while yearly the budget for the state is $70,000,000. What is the ratio of the budget for the city to that for the state?

Answers to Practice Problems: Ratios

1. $\frac{16}{9}$

2. $\frac{1}{3}$

3. $\frac{10}{3}$

4. $\frac{4}{3}$

5. $\frac{2}{3}$

6. $\frac{3}{10}$

7. $\frac{3}{1}$

8. $\frac{25}{1}$

9. $\frac{2}{1}$

10. $\frac{1}{2}$

11. $\frac{7}{10}$

12. $\frac{15}{14}$

13. $\frac{20}{26}$ or $\frac{30}{39}$ etc.

14. $\frac{4}{3}$

15. $\frac{\text{(C) cents}}{22}$

16. $\frac{5}{9}$

17. $\frac{16}{29}$

18. $\frac{21}{13}$

19. $\frac{41}{47}$

20. $\frac{3}{14}$

13 Proportion

When one ratio equals another ratio, you have a proportion.

Examples

a. $\frac{10}{5} = \frac{2}{1}$ **f.** $\frac{5}{2\frac{1}{2}} = \frac{2}{1}$

b. $\frac{9}{12} = \frac{3}{4}$ **g.** $\frac{6.5}{.5} = \frac{13}{1}$

c. $\frac{16}{30} = \frac{8}{15}$ **h.** $\frac{4\frac{1}{2}}{9} = \frac{1}{2}$

d. $\frac{27}{6} = \frac{9}{2}$ **i.** $\frac{135}{50} = \frac{27}{10}$

e. $\frac{124}{4} = \frac{31}{1}$ **j.** $\frac{.050}{.25} = \frac{1}{5}$

Each proportion has 4 terms: numerators and denominators are called terms.

$$\frac{\text{Term (1)}\ 8}{\text{Term (2)}\ 12} = \frac{2\ \text{Term (3)}}{3\ \text{Term (4)}}$$

Terms (1) and (4) are called the *extremes*.

Terms (2) and (3) are called the *means*.

LAW

In a proportion, the product of the means is equal to the product of the extremes.

Example

$$\text{Proportion } \frac{9}{15} = \frac{3}{5}$$

$$15 \times 3 \text{ (the means)} = 9 \times 5 \text{ (the extremes)}$$

$$\text{product } 45 = \text{product } 45$$

STUDY PROBLEM 1

Arnold drives 140 miles in 6 hours. Driving at the same rate, how long will it take him to make a trip of 210 miles?

Let $x =$ number of hours to make a trip of 210 miles

$$\text{Proportion: } \frac{140 \text{ miles}}{6 \text{ hours}} = \frac{210 \text{ miles}}{x \text{ hours}}$$

Using the law that the product of the means equals the product of the extremes, we have:

$$140(x) = 6(210)$$
$$140x = 1{,}260$$
$$x = 9 \text{ hours}$$

STUDY PROBLEM 2

A telephone pole 60 feet high casts a shadow 80 feet long at the same time that a tree casts a shadow 120 feet. What is the height of the tree?

Let $x =$ height of the tree

$$\text{Proportion: } \frac{60 \text{ ft}}{x} \frac{\text{height of pole}}{\text{height of tree}} = \frac{80 \text{ ft}}{120} \frac{\text{shadow of pole}}{\text{shadow of tree}}$$

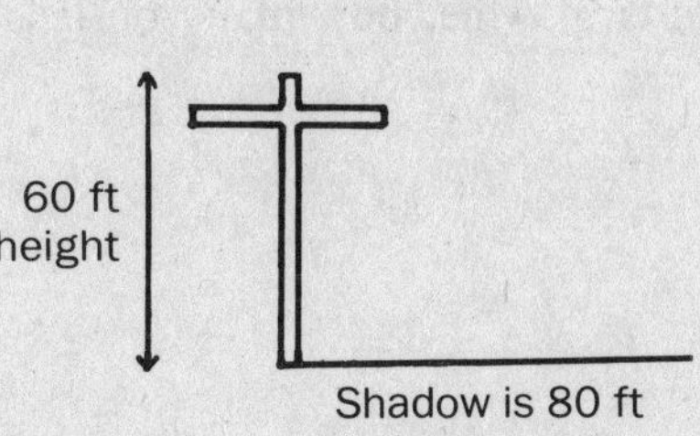

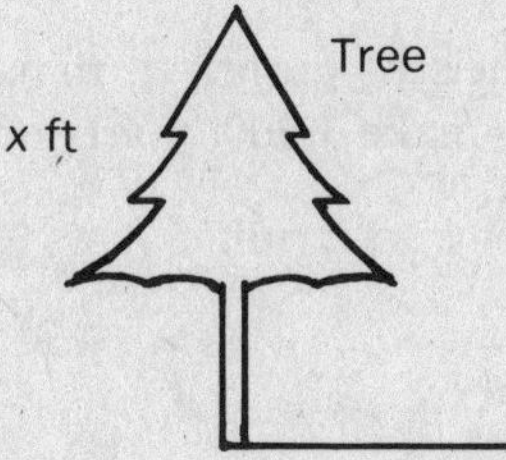

Law: Product of means equals product of extremes

$$80 \cdot x = 60 \cdot 120$$
$$80x = 7{,}200$$
$$x = 90 \text{ ft (height of tree)}$$

STUDY PROBLEM 3

If 18 gallons of gasoline cost \$23.31, how much would 12 gallons cost?

Let $x =$ cost for 12 gallons

$$\text{Proportion: } \frac{18}{23.31} \frac{\text{gal}}{\text{cost}} = \frac{12}{x} \frac{\text{gal}}{\text{cost}}$$

$$18(x) = 12(23.31)$$
$$18x = \$279.72$$
$$x = \$15.54$$

Practice Problems: Proportion

1. If 12 pencils cost 42 cents, how much would 100 pencils cost? Substitute your solutions for the question marks.

   Let $x = ?$

   Proportion: $\frac{?}{?} = \frac{?}{?}$

   Answer: ___?___

2. It takes a crew of 6 men to wash 84 windows of a hospital in one day. At the same rate, how many men would it take to wash 140 windows? Substitute your solution for the question marks.

   Let $x =$ ___?___

   Proportion: $\frac{?}{?} = \frac{?}{?}$

   Answer: ___?___

3. If it takes 10.5 bushels of grapes to make 30 quarts of wine, how many bushels of grapes will it take to make 3,200 quarts of wine?

   Let $x =$ ___?___

   Proportion: $\frac{?}{?} = \frac{?}{?}$

   Answer: ___?___

4. If 6 flowerpots cost $17.94, how much would 20 flowerpots cost?

   Let $x =$ cost of 20 flowerpots

   Proportion: $\frac{?}{?} = \frac{?}{?}$

   Answer: ___?___

5. A hotel 300 feet high casts a 240-foot shadow at the same time that a taller hotel casts a 360-foot shadow. How high is the taller hotel?

   Let $x =$ height of the taller hotel

   Proportion: $\frac{?}{?} = \frac{?}{?}$

   Answer: ___?___

6. A 6-foot tall man casts a shadow of 4 feet while a flagpole casts a shadow of 20 feet. How high is the flagpole?

Let x = height of flagpole

Proportion: $\frac{?}{?} = \frac{?}{?}$

Answer: ___?___

7. A woman died and left an estate valued at $220,000 to her grandson and to her son. If the ratio of the grandson's inheritance to the son's inheritance is 3 : 8, how much money did the son and grandson receive?

Let $3x$ = grandson's inheritance

Let $8x$ = son's inheritance

The sum of the two inheritances = ___?___

Equation is ___?___

Grandson received ___?___

Son received ___?___

Answers to Practice Problems: Proportion

1. x = cost of 100 pencils

$$\text{Proportion: } \frac{12}{.42} = \frac{100}{x}$$

$$12(x) = .42(100)$$

$$12x = 42.00$$

$$x = \$3.50 \text{ cost of 100 pencils}$$

2. Let x = number of men to wash 140 windows

$$\text{Proportion: } \frac{6 \text{ men}}{84 \text{ windows}} = \frac{x \text{ men}}{140 \text{ windows}}$$

$$84(x) = 6(140)$$

$$84x = 840$$

$$x = 10 \text{ men}$$

3. Let x = bushels of grapes to make 3,200 quarts of wine

$$\text{Proportion: } \frac{10.5 \text{ bushels}}{30 \text{ quarts}} = \frac{x \text{ bushels}}{3{,}200 \text{ quarts}}$$

$$10.5(3{,}200) = 30(x)$$

$$33{,}600 = 30x$$

$$1{,}120 = x \text{ bushels}$$

4.

$$\text{Proportion: } \frac{6}{17.94} = \frac{20}{x}$$

$$6 \cdot x = 20(17.94)$$

$$6x = 358.80$$

$$x = \$59.80 \text{ cost of 20 flowerpots}$$

5.

$$\text{Proportion: } \frac{300 \text{ ft}}{240} = \frac{x}{360}$$

$$240x = 300(360)$$

$$240x = 108{,}000$$

$$x = 450 \text{ ft height of taller hotel}$$

6.

$$\text{Proportion: } \frac{6}{4} = \frac{x}{20 \text{ ft}}$$

$$4x = 120 \text{ ft}$$

$$x = 30 \text{ ft height of flagpole}$$

7. $11x$ = sum of the two inheritances

Equation: $11x = \$220{,}000$

$x = \$\ 20{,}000$ one share

$3x = \$\ 60{,}000$ grandson's share

$8x = \$160{,}000$ son's share

14 Inequalities

Symbols:

a. $>$ is greater than

b. $<$ is less than

c. $=$ is equal to

d. $\geq$ is greater than or is equal to

e. $\leq$ is less than or is equal to

So far, we have solved equations where both sides of the equation were equal. In the equation $6x = 24$, for example, x is *equal* to 4.

Sometimes, however, both sides of an equation are *not equal.* In these cases, the solution to the equation may be a series or *set* of integers greater than or less than a given number.

Example 1

$$x + 3 < 7$$

In this example, x is any positive integer which, when added to 3, is *less* than 7. The solution, then, is not one number but a *set* of whole numbers.

The *solution set* for the equation above can be described by dots on a graph as follows.

0 1 2 3 4 5

Keep in mind that zero (0) is not a positive integer, therefore a dot is omitted. The above graph represents the solution set of $x + 3 < 7$. Solution Set: [1, 2, 3]

Example 2

$x + 3 > 7$ Solution Set: [5, 6, 7, ... *ad infinitum*]

−3 −2 −1 0 1 2 3 4 5 6 7

The above graph indicates that 4 is not a member of the solution set, but every point greater than 4 is a member of the solution set.

Example 3

$$-1 \leq x \leq 4$$

The solution set for the inequality above is in between −1 and +4.

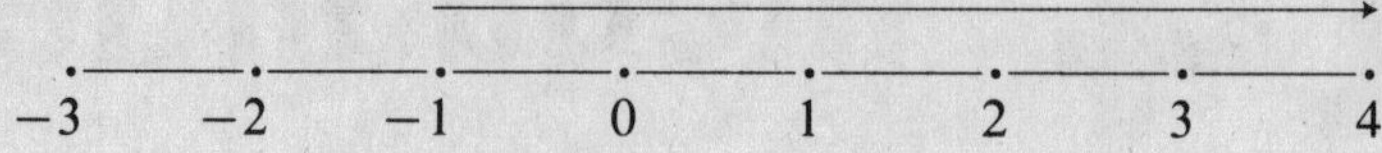

Example 4

a. If $x > 4$, the solution set is [5, 6, 7, 8, ... *ad infinitum*]. If we add 5 to each side of the inequality we have $x + 5 > 4 + 5$.

b. Then $x + 5 > 9$ [5, 6, 7, 8, ...]

Notice (a) and (b) have the same solution set, and this suggests that we have axioms for inequalities.

AXIOM 1

If the same number is added to both members of the inequality, the result is an equivalent inequality.

Example 5

a. If $x > 4$ Solution Set: [5, 6, 7, 8, ...]
Subtracting 2 from each side, we have $x - 2 > 4 - 2$.

b. Then, $x - 2 > 2$ Solution Set: [5, 6, 7, 8, ...]

Notice (a) and (b) have the same solution set.

AXIOM 2

If the same number is subtracted from both members of the inequality, the result is an equivalent inequality.

Example 6

a. If $x > 4$ Solution Set: [5, 6, 7, 8, ...]
Multiplying both sides of the inequality by 3, we have:

b. $3x > 12$ Solution Set: [5, 6, 7, 8, ...]

Notice (a) and (b) have the same solution set.

AXIOM 3

If both members of an inequality are multiplied by the same number, the result is an equivalent inequality.

Example 7

a. If $x > 4$ Solution Set: [5, 6, 7, 8, ...]

Dividing both sides of the inequality by 2, we have

$\frac{x}{2} > \frac{4}{2}$.

b. Then $\frac{x}{2} > 2$ Solution Set: [5, 6, 7, 8, ...]

Notice (a) and (b) have the same solution set.

AXIOM 4

If the same number is divided into each member of the inequality, the result is an equivalent inequality.

Example 8

a. If $x > 4$ Solution Set: [5, 6, 7, 8, ...]

Multiplying both sides by -3 we have:

b. $-3x > -12$ Solution Set: [0, 1, 2, 3]

BUT if you reverse the inequality to:

c. $-3x < -12$ Solution Set: [5, 6, 7, 8, ...]

Notice (a) and (c) are equal now.

AXIOM 5

If both members of an inequality are multiplied by a negative number, the inequality sign must be reversed.

Example 9

a. If $x > 4$ Solution Set: [5, 6, 7, 8, ...]

Dividing both sides by -2 we have

b. $\frac{x}{-2} > \frac{4}{-2}$ Solution Set: [0, 1, 2]

c. $\frac{x}{-2} < -2$ Solution Set: [5, 6, 7, 8, ...]

Notice (a) and (c) have the same solution set.

AXIOM 6

If both members of an inequality are divided by a negative number, the sign of the inequality is reversed.

Study the solutions provided for the inequalities that follow:

STUDY PROBLEM 1

$$4t+3>3t-7$$
$$4t-3t>-7-3$$
$$t>-10$$

STUDY PROBLEM 2

$$2x-5<-x+10$$
$$2x+x<+10+5$$
$$3x<15$$
$$x<5$$

STUDY PROBLEM 3

$$3(x+2)>-2(x+2)$$
$$3x+6>-2x-4$$
$$3x+2x>-4-6$$
$$5x>-10$$
$$x>-2$$

STUDY PROBLEM 4

$$-3(2m-8)<2(m+8)$$
$$-6m+24<2m+16$$
$$-6m-2m<+16-24$$
$$-8m<-8$$
$$m>+1$$

Dividing both sides by -8, the inequality sign is reversed.

STUDY PROBLEM 5

$$-3x+6(x-4)>10x+25$$
$$-3x+6x-24>10x+25$$
$$-3x+6x-10x>25+24$$

$$-7x > +49$$
$$x < -7$$

Dividing both sides by -7, the inequality sign is reversed.

STUDY PROBLEM 6

$$\frac{7m}{5} + \frac{13}{1} \leq \frac{3+m}{1}$$

$$(\not{5})\frac{7m}{\not{5}} + (5)\frac{13}{1} \leq \frac{(3+m)(5)}{1}$$

Steps:

1. L.C.D. = 5
2. Multiply each numerator by 5
3. Divide each numerator by the denominator
4. Clear fractions and solve

$$7m + 65 \leq 15 + 5m$$
$$7m - 5m \leq 15 - 65$$
$$2m \leq -50$$
$$m \leq -25$$

Practice Problems: Solving Inequalities

Solve each inequality.

1. $3t + 3 > 4t + 7$

2. $5 - 6x > 23$

3. $-7(3x + 4) > 35$

4. $3(y + 2) + 8 < 10 - (3y - 4)$

5. $-8(-2x - 3) + 5 \leq 5(3x - 5)$

6. $\frac{r}{4} - \frac{7r}{12} > 2$

7. $\frac{(a+2)}{5} + \frac{(a-3)}{2} \leq \frac{8}{1}$

Answers to Practice Problems: Solving Inequalities

1.

$$3t + 3 > 4t + 7$$
$$3t - 4t > 7 - 3$$
$$-t > 4$$
$$t < -4$$

2.

$$\begin{aligned} 5-6x &> 23 \\ -6x &> 23-5 \\ -6x &> 18 \\ x &< -3 \end{aligned}$$

3.

$$\begin{aligned} -7(3x+4) &> 35 \\ -21x-28 &> 35 \\ -21x &> 35+28 \\ -21x &> 63 \\ x &< -3 \end{aligned}$$

4.

$$\begin{aligned} 3(y+2)+8 &< 10-(3y-4) \\ 3y+6+8 &< 10-3y+4 \\ 3y+3y &< 10+4-8-6 \\ 6y &< 0 \\ y &< 0 \end{aligned}$$

5.

$$\begin{aligned} -8(-2x-3)+5 &\le 5(3x-5) \\ +16x+24+5 &\le 15x-25 \\ 16x-15x &\le -25-24-5 \\ 1x &\le -54 \end{aligned}$$

6.

$$\frac{r}{4}-\frac{7r}{12}>2$$

1. L.C.D. = 12
2. Multiply each numerator by the L.C.D.
3. Divide each numerator by the denominator of each fraction.
4. Solve the inequality.

$$\frac{{}^{3}(\cancel{12})\,r}{\cancel{4}}-\frac{(\cancel{12})\,7r}{\cancel{12}}>\frac{(12)\,2}{1}$$

$$\begin{aligned} 3r-7r &> 24 \\ -4r &> 24 \\ r &< -6 \end{aligned}$$

7.

$$\frac{{}^{2}(\cancel{10})\,(a+2)}{\cancel{5}}+\frac{{}^{5}(\cancel{10})\,(a-3)}{\cancel{2}}\le\frac{(10)\,8}{1}$$

1. L.C.D. = 10
2. Multiply each numerator by the L.C.D.
3. Divide each numerator by the denominator of each fraction.
4. Solve the inequality.

$$\begin{aligned} 2(a+2)+5(a-3) &\le 80 \\ 2a+4+5a-15 &\le 80 \end{aligned}$$

15 Solving Problems with Two Variables

In the previous chapters of this book, word problems were solved using one unknown, namely x, or any other single letter.

This chapter will show you how to solve word problems using two variables (x, y) instead of one. Using two variables gives you greater flexibility and allows you to solve more difficult problems.

Two variables means that we must establish two equations. The standard format for equations with two unknowns is as follows:

> Both the x and y variables are placed in the left member. The constants are placed in the right member.

Examples

a. $x = y + 10 \longrightarrow x - y = 10$

b. $2y = -2x - 15 \longrightarrow 2x + 2y = -15$

c. $2(y - x) = -3y + 16$

$2y - 2x = -3y + 16 \longrightarrow -2x + 5y = 16$

d. $6(-x - y) = 3(3x + 8)$

$-6x - 6y = 9x + 24$

$-6x - 9x - 6y = 24 \longrightarrow -15x - 6y = 24$

e. $\frac{3y}{7} - \frac{2x}{21} = 1$

L.C.D. = 21

$9y - 2x = 21 \longrightarrow -2x + 9y = 21$

f. $-x - 6 = -y - 6$

$-x + y = -6 + 6 \longrightarrow -x + y = 0$

g. $\frac{(2 - x)}{2} + \frac{(y + 3)}{4} = \frac{3(y - 6)}{3}$

L.C.D. = 12

$$7a - 11 \le 80$$
$$7a \le 91$$
$$a \le 13$$

WORD PROBLEMS WITH INEQUALITIES

STUDY PROBLEM 8

A man in retirement found that he needed more than \$2,500 per year return on his investments to live comfortably. If he invested \$10,000 at 7%, how much must he invest at 9% to attain his goal?

Let x = amount invested at 9%
$.09x$ = return from 9% investment
$.07(10{,}000)$ = return from 7% investment

Equation: 7% investment + 9% investment > 2,500

$$.09x + .07(10{,}000) > 2{,}500$$
$$9x + 70{,}000 > 250{,}000$$
$$9x > 180{,}000$$
$$x > 20{,}000$$

Practice Problems: Word Problems with Inequalities

1. The longer side of a triangle is 3 times the shorter side. The third side is 4 inches longer than the shorter. How long is the shorter side if the perimeter is more than 39 inches?

2. The sum of 5 times a number and 50 is less than 6 times the number. What numbers satisfy this inequality?

3. A teenager buys a sports jacket and a pair of jogging shorts. He decides to spend no more than \$27 for both articles. If the sports jacket costs \$6 more than twice the jogging shorts, how much did he spend for the jogging shorts?

4. Deena weighs 12 pounds less than her older sister Julia. Their combined weights total less than 200 pounds. What inequality expression satisfies Julia's weight?

5. A certain type of coolant contains a 65% efficiency rating against heat. Another type contains a 90% efficiency rating against heat. How many quarts of the 90% and 65% efficiency rating must be mixed together to get at least 8 quarts with an 80% efficiency rating?

Answers to Practice Problems: Word Problems with Inequalities

1. Let $x =$ inches in shorter side
$3x =$ inches in longer side
$x + 4 =$ inches in the third side

Equation: $x + 3x + x + 4 > 39''$
$5x + 4 > 39''$
$5x > 35$
$x > 7$ inches on the shorter side

2. Let $x =$ the number

Equation: $5x + 50 < 6x$
$5x - 6x < -50$
$-x < -50$
$x > 50$

3. Let $x =$ cost of jogging shorts
$2x + 6 =$ cost of sports jacket

Equation: $x + 2x + 6 \leq 27$
$3x \leq 27 - 6$
$3x \leq 21$
$x \leq \$7$

4. Let $x =$ Julia's weight
$x - 12 =$ Deena's weight

Equation: $x + x - 12 < 200$ lb
$2x - 12 < 200$ lb
$2x < 200 + 12$
$2x < 212$
$x < 106$ lb

5.

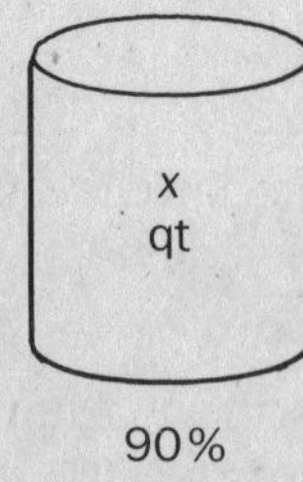

90%

+

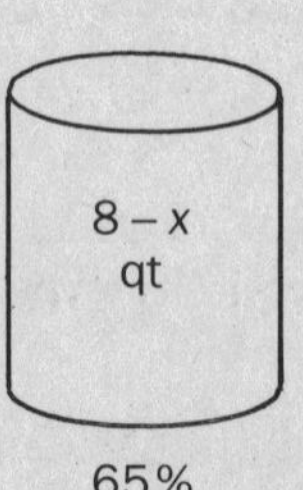

65%

≥

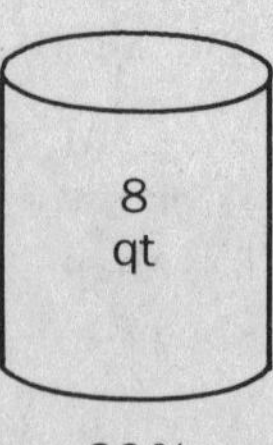

80%

Equation: $.90(x) + .65(8 - x) \geq .80(8)$

Move decimal point 2 places to the right.

$90x + 520 - 65x \geq 640$
$90x - 65x \geq 640 - 520$
$25x \geq 120$
$25x \geq 120$ qt
$x \geq 4\frac{4}{5}$ qt of the 90% rating
$8 - x \geq 3\frac{1}{5}$ qt of the 65% rating

$$6(2-x)+3(y+3)=12(y-6)$$
$$12-6x+3y+9=12y-72$$
$$-6x+3y-12y=-72-21 \longrightarrow -6x-9y=-93$$

Practice Problems: Two Variables (Setting Up the Equation)

1. $x=-y+16$
2. $3(-4x+2)=-y$
3. $-8(x+y)=-3(x-y)+8$
4. $\frac{-6x}{9}-\frac{3y}{2}=\frac{3x+5}{3}$
5. $\frac{(x+y)}{6}-\frac{x}{12}=\frac{3(-x-y)}{6}$

Answers to Practice Problems: Two Variables (Setting Up the Equation)

1. $x+y=16$
2. $-12x+y=-6$
3. $-5x-11y=8$
4. $-30x-27y=30$
5. $7x+8y=0$

STUDY PROBLEM 1

The sum of two numbers is 15 and their difference is 7. What are the two numbers?

Let $x=$ larger number
$y=$ smaller number

Equations: (1) $x+y=15$
(2) $x-y=7$

By adding both equations (1) and (2) we have

$$2x=22$$
$$x=11 \text{ (larger number)}$$

Substituting 11 for x in equation (1) we have

$$11+y=15$$
$$+y=15-11$$
$$y=4 \text{ (smaller number)}$$

Solution Set (11, 4)

RULE

To eliminate the same variable of both equations, the signs of each must be the inverse of the other and their coefficients must be the same.

STUDY PROBLEM 2

$$\text{Equation (1)} \quad x+y=8$$
$$\text{Equation (2)} \quad x-2y=2$$

Notice that the x variables in both equations have the same coefficients (1) and the same sign (+). If we change the sign of either equation (1) or equation (2), the x variable will be eliminated. Here we can use the subtraction method.

Subtraction method:

$$\begin{array}{rr} (1) & -x-y=-8 \\ (2) & \underline{x-2y=2} \\ & -3y=-6 \\ & y=2 \end{array}$$

Notice that if we reverse the sign of x *to* −x, *we must also reverse the signs of the other terms in the equation.*

Substituting 2 for y in equation (2), we have

$$x-4=2$$
$$x=6$$

Solution Set (6, 2)

STUDY PROBLEM 3

$$\text{Equation (1)} \quad x+2y=8$$
$$\text{Equation (2)} \quad -2x+2y=-10$$

Using subtraction method

$$x=6$$

Substituting 6 for x in equation (1)

$$6+2y=8$$
$$2y=2$$
$$y=1$$

Solution Set (6, 1)

STUDY PROBLEM 4

Caren's petty cash contains 100 coins made up of nickels and quarters. How many nickels and quarters does she have in her petty cash if the worth of all coins is $17.00?

Let x = number of quarters
Let y = number of nickels

Equation (1) $x+y=100$ coins
Equation (2) $.25x+.05y=17.00$

Move decimal point 2 places to the right.

(1) $x+y=100$
(2) $25x+5y=1{,}700$

Multiplying equation (1) by -5

(1) $-5x-5y=-500$
(2) $25x+5y=1{,}700$

Adding (1) and (2)

$20x=1{,}200$
$x=60$ quarters

$y=40$ nickels

As you can see above, there are *three methods* for eliminating the same variable of both equations:

1. addition
2. subtraction
3. multiplication

Practice Problems: Two Variables (Addition and Subtraction Methods)

Using the addition method, solve for x and y in Problems 1 through 4.

1. (1) $x+y=12$
(2) $x-y=4$

2. (1) $6x-5y=9$
(2) $2x+5y=23$

3. (1) $-8x+3y=-5$
(2) $8x-2y=6$

4. (1) $.5x+.7y=6.4$
(2) $.2x-.7y=.6$

Place in standard format and then use the addition method to solve Problem 5.

5. (1) $3(x+1)=2(y+1)$
(2) $2y+3=2(x+4)$

Using the subtraction method, solve for x and y in Problems 6 through 8.

6. (1) $x+2y=16$
(2) $2x+2y=24$

7. (1) $-x+3y=6$
(2) $-x+y=10$

8. (1) $.08x-.05y=6$
(2) $.08x-.02y=3$

Rearrange to standard format and then use the subtraction method to solve Problem 9.

9. (1) $3(x+2y)=-2x+8$
(2) $2(y+3)=x-4y-10$

Answers to Practice Problems: Two Variables (Addition and Subtraction Methods)

Solution sets for the addition method.

1. (8, 4)

2. (4, 3)

3. (1, 1)

4. (10, 2)

5. (4, 6½)

Solution sets for the subtraction method.

6. (8, 4)

7. (−12, −2)

8. (12½, −100)

9. (4, −2)

STUDY PROBLEM 5

Use the multiplication method to eliminate the same variable of each equation. Then solve for x and y.

$$(1)\quad x+2y=-1$$
$$(2)\quad 5x-4y=16$$

Multiplying equation (1) by 2 we get

$$\begin{aligned}(1)\quad 2x+4y&=-2\\(2)\quad \underline{5x-4y}&\underline{=16}\\7x\qquad&=14\\x&=2\end{aligned}$$

Substituting 2 for x in equation (1)

$$\begin{aligned}(1)\quad 2+2y&=-1\\2y&=-1-2\\2y&=-3\\y&=-3/2=-1\tfrac{1}{2}\end{aligned}$$

Solution Set is (2, −1½)

STUDY PROBLEM 6

Use the multiplication method to eliminate the same variable of each equation. Then solve for x and y.

$$(1) \quad 4x+2y=0$$
$$(2) \quad x-5y=-11$$

Multiplying equation (2) by -4 we get

$$(1) \quad 4x+2y=0$$
$$(2) \quad -4x+20y=+44$$
$$22y=+44$$
$$y=+2$$

Substituting 2 for y in equation (1)

$$(1) \quad 4x+4=0$$
$$4x=-4$$
$$x=-1$$

Solution Set is $(-1, 2)$

STUDY PROBLEM 7

Use the multiplication method to eliminate the same variable of each equation. Then solve for x and y.

$$(1) \quad 3x-8y=-18$$
$$(2) \quad 2x+5y=19$$

Multiplying equation (1) by -2 and equation (2) by $+3$ we get

$$(1) \quad -6x+16y=36$$
$$(2) \quad 6x+15y=57$$
$$31y=93$$
$$y=3$$

Substituting 3 for y in equation (1)

$$(1) \quad 3x-24=-18$$
$$3x=24-18$$
$$3x=6$$
$$x=2$$

Solution Set is $(2, 3)$

Practice Problems: Two Variables (Multiplication Method)

Using the multiplication method, solve for x and y in the following problems.

1. (1) $2x+3y=23$
   (2) $3x+y=17$

2. (1) $5x+y=26$
   (2) $x+5y=-14$

3. (1) $5x-3y=23$
   (2) $7x-4y=33$

Rearrange to standard format and find the solution set.

4. (1) $2x+3y=-5x-2y+31$
   (2) $-2(7x-2y)=-2(y+15)$

5. (1) $-1-3y=2(y-2x)$
   (2) $2(x-y)=2$

Answers to Practice Problems: Two Variables (Multiplication Method)

1.

$$\begin{aligned} (1) \quad 2x+3y&=23 \\ (2) \quad 3x+y&=17 \end{aligned}$$

Multiplying (2) by -3

$$\begin{aligned} (1) \quad -9x-3y&=-51 \\ 2x+3y&=23 \\ \hline -7x&=-28 \\ x&=+4 \end{aligned}$$

Substituting 4 for x in (1)

$$\begin{aligned} 8+3y&=23 \\ 3y&=23-8 \\ 3y&=15 \\ y&=5 \end{aligned}$$

Solution Set (4, 5)

2.

$$\begin{aligned} (1) \quad 5x+y&=26 \\ (2) \quad x+5y&=-14 \end{aligned}$$

Multiplying (2) by -5

$$\begin{aligned} (1) \quad -5x-25y&=+70 \\ 5x+y&=26 \\ \hline -24y&=96 \\ y&=-4 \end{aligned}$$

Substituting -4 for y in (1)

$$\begin{aligned} 5x-4&=26 \\ 5x&=30 \\ x&=6 \end{aligned}$$

Solution Set (6, −4)

3.

$$(1) \quad 5x-3y=23$$
$$(2) \quad 7x-4y=33$$

Multiplying (1) by −4 and (2) by 3

$$\begin{array}{rr} (1) & -20x+12y=-92 \\ (2) & 21x-12y=99 \\ \hline & 1x=7 \end{array}$$

Substituting 7 for x in (1)

$$\begin{array}{rr} (1) & 35-3y=23 \\ & -3y=-12 \\ & y=+4 \end{array}$$

Solution Set (7, 4)

4. Rearranged

$$(1) \quad 7x+5y=31$$
$$(2) \quad -14x+6y=-30$$

Multiplying (1) by 2

$$\begin{array}{r} 14x+10y=62 \\ -14x+6y=-30 \\ \hline 16y=32 \\ y=2 \end{array}$$

Substituting 2 for y in (1)

$$\begin{array}{r} 7x+10=31 \\ 7x=21 \\ x=3 \end{array}$$

Solution Set (3, 2)

5. Rearranged

$$(1) \quad 4x-5y=1$$
$$(2) \quad 2x-2y=2$$

Multiplying (2) by −2

$$\begin{array}{rr} (2) & -4x+4y=-4 \\ & 4x-5y=1 \\ \hline & -1y=-3 \\ & y=+3 \end{array}$$

Substituting 3 for y in (1)

$$\begin{array}{r} 4x-15=1 \\ 4x=16 \\ x=4 \end{array}$$

Solution Set (4, 3)

FRACTIONAL EQUATIONS CONTAINING TWO VARIABLES

When one or both of the equations to be solved contains a fraction, you must find the lowest common denominator and clear the fractions before eliminating one variable and solving for x and y.

STUDY PROBLEM 8

(1) $$\frac{x}{2}+\frac{y}{3}=2$$

(2) $$\frac{x}{3}+\frac{y}{9}=1$$

Equation (1) L.C.D. $=6$ (multiply each fraction by 6)

$$6\left(\frac{x}{2}\right)+6\left(\frac{y}{3}\right)=6\left(\frac{2}{1}\right)$$

$$3x+2y=12$$

Equation (2) L.C.D. $=9$ (multiply each fraction by 9)

$$9\left(\frac{x}{3}\right)+9\left(\frac{y}{9}\right)=9\left(\frac{1}{1}\right)$$

$$3x+y=9$$

Subtract Equation (1) from Equation (2)

$$\begin{array}{r} 3x+y=9 \\ 3x+2y=12 \\ \hline -y=-3 \\ y=+3 \end{array}$$

Substituting $(+3)$ for y in equation (2)

(2) $$\begin{aligned} 3x+3&=9 \\ 3x&=6 \\ x&=2 \end{aligned}$$

Solution Set $(2, 3)$

Practice Problems: Two Variables (Any Method)

In the systems of equations on page 101, use any of the three convenient methods to find the solution set.

1. (1) $3x-5y=4$
   (2) $2x+5y=16$
2. (1) $-7x+y=2$
   (2) $7x+2y=4$
3. (1) $-3x+2y=6$
   (2) $-4x-2y=-20$
4. (1) $2x+3y=4$
   (2) $x+3y=14$
5. (1) $x=3y-4$
   (2) $-x=2y-16$
6. (1) $11x-y=17$
   (2) $7x-4y=-6$
7. (1) $-q=-8p+22$
   (2) $7p=q+18$
8. (1) $4x-3y=-5$
   (2) $2x-5y=22$
9. (1) $2(x+y)+10=5(x+1)$
   (2) $x+y-2=2(y-1)$
10. (1) $\frac{x}{2}-\frac{3y}{4}=-2$
    (2) $\frac{y}{2}+\frac{x}{4}=\frac{4}{1}$

Check your answers below.

Answers to Practice Problems: Two Variables

1. $(4, 1\frac{3}{5})$
2. $(0, 2)$
3. $(2, 6)$
4. $(-10, 8)$
5. $(8, 4)$
6. $(2, 5)$
7. $(4, 10)$
8. $(-6\frac{1}{2}, -7)$
9. $(5, 5)$
10. $(4\frac{4}{7}, 5\frac{5}{7})$

WORD PROBLEMS CONTAINING TWO VARIABLES

STUDY PROBLEM 9

Bob wants to create a special blend of tea by mixing tea costing \$1.65 per pound with tea costing \$1.85 per pound. To make up 7 pounds of the special blend at a cost of \$1.75 per pound, how many pounds of each brand of tea should he use?

Model:

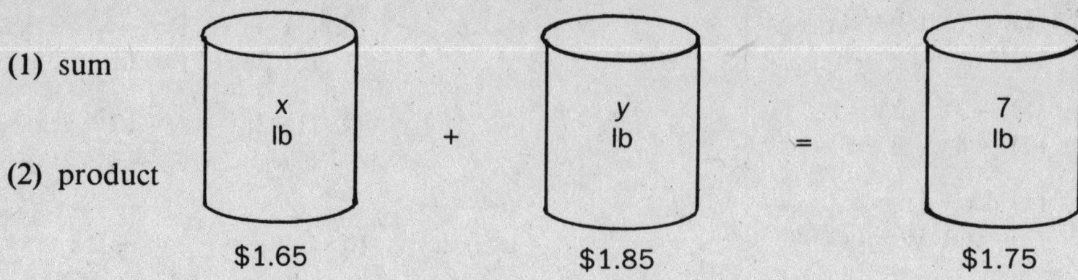

Since we have *two* unknowns, x and y, we must generate *two* equations. These equations we will call (1) *sum* and (2) *product*. The sum equation deals with pounds.

$$(1) \qquad x \text{ lb} + y \text{ lb} = 7 \text{ lb}$$

The product deals with cost.

$$(2) \quad \$1.65x + \$1.85y = \$12.25$$

Equation (1) $\quad x + y = 7$

Equation (2) $\quad 100(1.65x + 1.85y) = 100(12.25)$

Multiplying by 100 moves the decimal point 2 places so that we now have whole numbers.

$$165x + 185y = 1{,}225$$

Equation (1) $\quad x + y = 7$

Equation (2) $\quad 165x + 185y = 1{,}225$

Our next step is to eliminate the x's by multiplying the equation (1) by -165.

$$-165(x + y) = -165 \cdot 7$$

Equation (1) $\quad -165x - 165y = -1{,}155$

Equation (2) $\quad 165x + 185y = 1{,}225$

By addition $\quad 20y = 70$

$$y = 3\tfrac{1}{2} \text{ lb}$$

$$x + 3\tfrac{1}{2} = 7$$

Substituting in Equation (1) $\quad x = 3\tfrac{1}{2} \text{ lb}$

STUDY PROBLEM 10

Enrico sold a 10-pound box of assorted candy. Some of the candy in the box sold for \$1.75 per pound and some sold for \$2.00 per pound. How many pounds of each variety were included in the box if the mixture sold for \$1.90 per pound?

Let $x =$ lb of \$1.75 variety
$y =$ lb of \$2.00 variety

Model:

(1) sum

(2) product

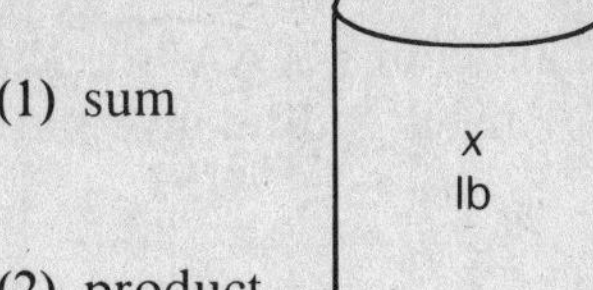
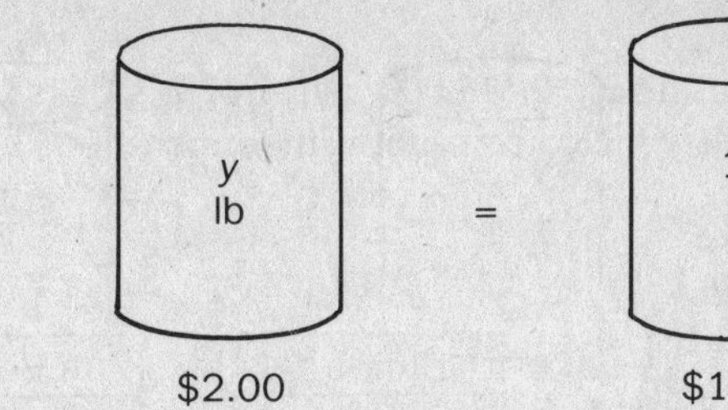

Sum equation (1) $x+y=10$ lb

Product equation (2) $1.75x+2.00y=19.00$

Multiplying by 100 moves the decimal point 2 places. Multiply equation (1) by -200.

$$\begin{aligned}
(1)\quad & -200(x-y)=-200\cdot 10 \\
(1)\quad & -200x-200y=-2{,}000 \\
(2)\quad & \underline{175x+200y=1{,}900} \\
& -25x=-100 \\
& x=4 \text{ lb @ \$1.75/lb}
\end{aligned}$$

Substituting in Equation (1) $\quad y=6$ lb @ \$2.00/lb

STUDY PROBLEM 11

Susan's piggy bank contains 96 coins made up of nickels and quarters. How many nickels and quarters does she have in the piggy bank if the worth of all coins is \$12.00?

Let $x=$ number of nickels
Let $y=$ number of quarters

Sum equation (1) $x+y=96$ (number of coins)

Let $5x=$ value of the nickels in pennies
Let $25y=$ value of the quarters in pennies

Product equation (2) $5x+25y=1{,}200$ (total value of coins in pennies)

Multiplying equation (1) by -5

$$\begin{aligned}
(1)\quad & -5x-5y=-480 \\
(2)\quad & \underline{5x+25y=1{,}200}
\end{aligned}$$

By addition $\quad 20y=720$

$y=36$ quarters

Substituting in Equation (1) $\quad x=60$ nickels

STUDY PROBLEM 12

A man has invested $15,000, part in 7% stocks and the remainder in 9% AA Bonds. If his yearly yield from both investments is $1,230.00, how much did he invest at each rate?

Model:

	Stocks	+	AA Bonds	=	15,000 Invested
(1) sum	x		y		Annual Yield
(2) product	7%		9%		$1,230.00

Sum equation (1) $x+y=\$15,000$

Product equation (2) $7\%x+9\%y=\$1,230.00$

Multiplying (2) by 100

(2) $.07x+.09y=\$1,230.00$

(2) $7x+9y=123,000$

(1) $x+y=15,000$

Multiplying (1) by -7

(1) $-7x-7y=-105,000$

(2) $7x+9y=\ \ 123,000$

Addition $2y=18,000$

$y=\$9,000$ invested in 9% AA Bonds

Substituting in (1) $x=\$6,000$ invested in 7% stocks

STUDY PROBLEM 13

A farmer wishes to combine feed that is 10% corn with feed that is 15% soybean to make a mixture that is 12% corn and soybean. If he makes up 1,200 pounds of the mixture, how many pounds of the corn mixture and how many pounds of the soybean mixture must he use?

Model:

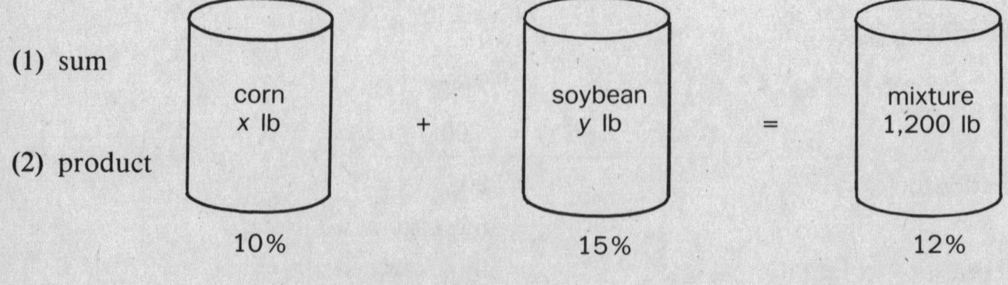

Sum equation (1) $x + y = 1{,}200$

Product equation (2) $10\%x + 15\%y = 12\%(1{,}200)$

Multiplying (2) by 100

(2) $.10x + .15y = .12(1{,}200)$

(2) $10x + 15y = 14{,}400$

(1) $x + y = 1{,}200$

Multiplying equation (1) by -10

(1) $-10x - 10y = -12{,}000$

(2) $10x + 15y = 14{,}400$

Adding $5y = 2{,}400$

$y = 480$ lb soybean mixture

Substituting in (1) $x = 720$ lb corn mixture

STUDY PROBLEM 14

A pharmacist has a 12% solution of boric acid and a 20% solution of boric acid. How many grams of the 12% and 20% solution must he use to make 100 grams of a 15% solution?

Model:

(1) sum

(2) product

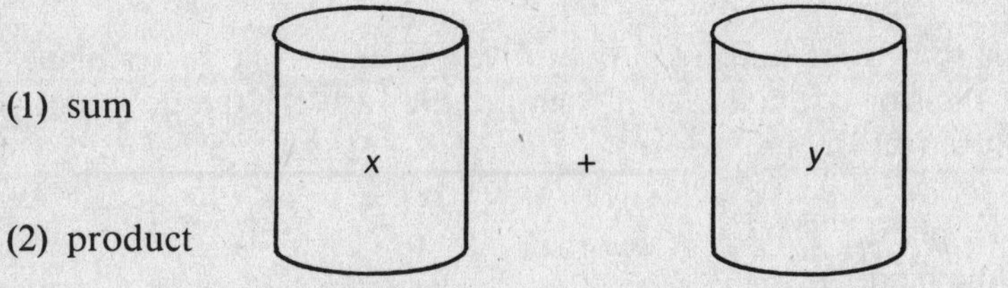

Sum equation (1) $x + y = 100$

Product equation (2) $.12x + .20y = .15(100)$

$12x + 20y = 1{,}500$

Multiplying (1) by -12 we get

(1) $-12x - 12y = -1{,}200$

(2) $12x + 20y = 1{,}500$

Add (1) and (2) $8y = 300$

$y = 37\frac{1}{2}$ grams of 20% solution

Substituting in (1) $x = 62\frac{1}{2}$ grams of 12% solution

STUDY PROBLEM 15

An airplane makes a test run with a tail wind and records a speed of 682 miles per hour. On the return trip against the same wind, the plane can attain only 624 miles per hour. What is the speed of the plane in still air and what is the wind's velocity?

Let $x =$ speed of plane in still air
Let $y =$ wind's velocity

With tail wind →

Equation (1) $x + y = 682$

← Return trip against wind

Equation (2) $x - y = 624$

By addition of (1) and (2)

$$2x = 1{,}306$$

$x = 653$ (speed of plane)

By substitution $y = 29$ (wind's velocity)

STUDY PROBLEM 16

Two planes start at the same time and fly in opposite directions. The faster plane travels twice as fast as the slower plane. After 3 hours, they are 2,520 miles apart. How fast does each plane travel?

Let $x =$ speed of the fast plane
Let $y =$ speed of the slow plane

Equation (1) $x = 2y$
Equation (2) $3x + 3y = 2{,}520$

(1) $x - 2y = 0$
(2) $3x + 3y = 2{,}520$

Multiply by -3

(1) $-3x + 6y = 0$
(2) $3x + 3y = 2{,}520$

By addition $9y = 2{,}520$

$y = 280$ mph (slower plane)

By substitution $x = 560$ mph (faster plane)

STUDY PROBLEM 17

Four hundred tickets were sold for a school play, bringing in a total of $1,025.00. If adults were charged $3.50 and students $2.00, how many adults and how many students attended the performance?

Model:

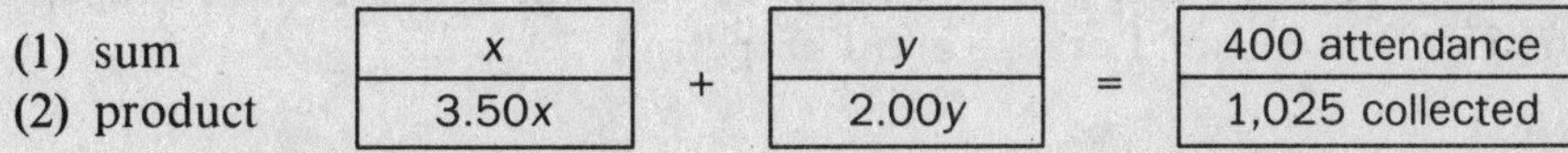

(1) sum	x	+	y	=	400 attendance
(2) product	$3.50x$		$2.00y$		1,025 collected

Let $x =$ adult tickets sold
Let $y =$ student tickets sold

(1) $x + y = 400$

(2) $3.50x + 2.00y = 1{,}025.00$

(2) $350x + 200y = 102{,}500$

Multiplying (1) by -200

(1) $-200x - 200y = -80{,}000$

By addition (2) $150x = 22{,}500$

$x = 150$ adult tickets

By substitution $y = 250$ student tickets

Practice Problems: Word Problems Using Two Variables

Check your answers with the solutions on pages 110–116.

1. The sum of two numbers is 48 and their difference is 12. What are the two numbers?

2. The sum of two numbers is 18. Twice the first less the second equals 6. What are the two numbers?

3. The sum of two numbers is 102 and their difference is 68. What are the numbers?

4. The sum of 2 numbers is 18. Three times the larger less the smaller equals 22. What are the numbers?

5. The perimeter of a rectangle is 120 cm. Three times its width equals the length. What are the dimensions?

6. There are 24 coins made up of quarters and dimes which amount to $3.90. How many of each kind of coin are there?

7. Mr. Owens, a financial consultant, recommended that a client invest $15,000, part at 7% and the remainder at 9%, to insure a $1,300.00 yearly income. How much was invested at each rate?

8. Two sodas and 2 sundaes cost $4.00. Three sodas and 1 sundae cost $3.90. What is the cost of 1 soda and of 1 sundae?

9. One dozen oranges and 6 grapefruit cost $2.46. Nine oranges and 9 grapefruit cost $2.97. Find the cost of 1 orange and 1 grapefruit.

10. Euclid High School received a shipment of algebra and geometry books costing $4,200. The cost of 1 algebra book is $12.00 and the cost of 1 geometry book is $15.00. If there were 125 more algebra books than geometry books, how many of each kind of math book were in the shipment?

11. A merchant has two kinds of Christmas candy, one selling at $2.20 a pound and the other at $.95 a pound. How many pounds of each should he mix to sell 100 pounds at $2.10?

12. Mr. Szuki died leaving an insurance policy amounting to $40,000 to his wife and son. His wife was to receive $2,000 less than 3 times what her son was to receive. What amount did the mother and son each receive?

13. A labor union has $100,000 to invest for its membership. It was decided that the investment would be in an 8% preferred stock and a 12% AA Bond. If the yearly income from both investments is $10,400, how much was invested in the stocks and in the AA Bonds?

14. If 1 pound of coffee and 8 pounds of sugar cost $9.11, and 5 pounds of coffee and 5 pounds of sugar cost $14.40, find the cost of 1 pound of coffee.

15. If 3 pizzas and 6 colas cost $13.95 and 2 pizzas and 5 colas cost $9.65, find the cost of 1 pizza and 1 cola.

16. Ralph Peterson, a graduating senior at U.S.C., received an annuity amounting to $14,800 from his grandparents. He quickly invested part of it in Treasury Bonds giving a 7% return and the remainder in preferred stocks yielding a 6½% yearly return. If he expects $1,002.00 from both investments, what amount did he invest in the Treasury Bonds and what amount in preferred stocks?

17. If Darlene can purchase 6 cans of peaches and 12 cans of tomatoes for $11.10 or 12 cans of peaches and 6 cans of tomatoes for $10.86, what is the price of 1 can of peaches and 1 can of tomatoes?

18. Mr. Blossom, a farmer, paid 10 men and 8 boys $970 for 1 day's work. Later he paid 12 men and 6 boys $1,020 for a day's work. The men were paid at a uniform wage and the boys were paid at another uniform wage. What daily wage was paid to the men and boys?

19. A sum of \$46,000 is invested, part in a small business enterprise which yields a 20% return and part in mutual funds which yield an 8% return. Find the amount invested in each part if the yearly yield is \$5,600.

20. A trust fund containing \$30,000 had been established for a young adult. Part of the money was invested in a Savings and Loan Association which paid a 5½% yearly yield and the remainder was invested in Second Trust Deeds yielding a 7% return. If he received \$1,830 from the two investments annually, how much did he invest in each?

21. A 40-pound weight is placed at one end of a 10-foot lever and a 10-pound weight is placed at the other end of the lever. At what distance from the 40-pound and 10-pound weight should the fulcrum be placed?

Hint: Let x = distance from fulcrum for the 40-lb weight
Let y = distance from the fulcrum for the 10-lb weight

weight 40 lb x y 10 lb weight
fulcrum

Equation (1) ______________________

Equation (2) ______________________

22. In a town in Kansas, 6,000 people attended a street fair held to raise funds for scholarships for needy students. Entrance fees were \$1.50 for children and 3 dollars for adults. How many adults and children attended if \$14,250 was collected?

23. Two cars started from the same point but traveled in opposite directions. The faster car traveled 15 kilometers faster than the slower car. After five hours the 2 cars were 575 kilometers apart. How many kilometers per hour did each car travel?

24. A softball game was planned by the Rotary Club to raise funds to build an athletic facility for handicapped persons. Adults other than senior citizens were charged 5 dollars. All others were charged 2 dollars each. If the total attendance was 3,702 and if the sum of 16,404 dollars was collected, how many adult tickets and all other tickets were sold?

25. An F-18 fighter pilot makes a test run from Andrews Air Force Base in California to La Guardia Airport in New York City averaging 770 miles per hour with a tail wind. On the return trip against the same wind, the plane averaged 680 miles per hour. What is the speed of the plane in still air and what is the velocity of the wind?

26. If 2 high school football players with a combined weight of 396 pounds balance a lever when they sit respectively 6 feet and 5 feet from the fulcrum, what is each player's weight?

27. At a famous university the alumni of the School of Liberal Arts and the School of Engineering decided they would together raise \$545,000 for their Alma Mater to commemorate the 25th anniversary of their graduation. There were a total of 4,200 pledges.

The Liberal Arts school averaged \$150.00 per pledge and the School of Engineering averaged \$100.00 per pledge. How many alumni from each school made pledges?

28. The average of 2 test scores is 83.50. If the difference between the scores is 5, what were the 2 test scores?

29. The sum of Ray's age and his sister Rosalie's age is 36. One-half the sum of Ray's age in 4 years is equal to ½ of twice Rosalie's age less 1. What are their present ages?

Hint: Let $x =$ Ray's age
Let $y =$ Rosalie's age
Find L.C.D. and clear equations

30. One-fifth of the sum of Lorna's age in 5 years is equal to ¼ of Larry's age less 1. Two-fifths of the sum of Larry's age in 4 years is equal to ⅕ of twice Lorna's age plus 2. What are their ages now?

Hint: Let $x =$ Lorna's age
Let $y =$ Larry's age
Find L.C.D. of each equation and then clear fractions

Answers to Practice Problems: Word Problems Using Two Variables

1.

$$\begin{array}{r} x+y=48 \\ x-y=12 \\ \hline 2x=60 \\ x=30 \\ y=18 \end{array}$$

2.

$$\begin{array}{r} x+y=18 \\ 2x-y=6 \\ \hline 3x=24 \\ x=8 \\ y=10 \end{array}$$

3.

$$\begin{array}{r} x+y=102 \\ x-y=68 \\ \hline 2x=170 \\ x=85 \\ y=17 \end{array}$$

4.

$$\begin{array}{r} x+y=18 \\ 3x-y=22 \\ \hline 4x=40 \\ x=10 \\ y=8 \end{array}$$

5.

$$2L + 2W = 120$$
$$3W = L$$

(Multiply by 2)

$$2L + 2W = 120$$
$$-L + 3W = 0$$

$$2L + 2W = 120$$
$$\underline{-2L + 6W = 0}$$
$$8W = 120$$
$$W = 15$$
$$L = 45$$

6. (Multiply by -10)

$$q + d = 24$$
$$25q + 10d = 390$$

$$-10q - 10d = -240$$
$$\underline{25q + 10d = 390}$$
$$15q = 150$$
$$q = 10$$
$$d = 14$$

7. (Multiply by -7)

$$x + y = 15{,}000$$
$$7x + 9y = 130{,}000$$

$$-7x - 7y = -105{,}000$$
$$\underline{7x + 9y = 130{,}000}$$
$$2y = 25{,}000$$
$$y = 12{,}500 \text{ @ } 9\%$$
$$x = 2{,}500 \text{ @ } 7\%$$

8.
(Multiply by -2)

$$2x + 2y = \$4.00$$
$$3x + 1y = \$3.90$$

$$2x + 2y = \$4.00$$
$$\underline{-6x - 2y = -\$7.80}$$
$$-4x = -\$3.80$$
$$x = \$.95 \text{ cost of 1 soda}$$
$$y = \$1.05 \text{ cost of 1 sundae}$$

9. (Multiply by -3)
(Multiply by 2)

$$12x + 6y = \$2.46$$
$$9x + 9y = \$2.97$$

$$-36x - 18y = -\$7.38$$
$$\underline{18x + 18y = \$5.94}$$
$$-18x = -\$1.44$$
$$x = 8 \text{ cents cost of 1 orange}$$
$$y = 25 \text{ cents cost of 1 grapefruit}$$

10.

(Multiply by -12)

$$12x+15y=4{,}200$$
$$x-y=125$$

$$12x+15y=4{,}200$$
$$-12x+12y=-1{,}500$$
$$27y=2{,}700$$
$y=100$ geometry books
$x=225$ algebra books

11. (Multiply by -95)

$$x+y=100$$
$$95x+220y=21{,}000$$

$$-95x-95y=-9{,}500$$
$$95x+220y=21{,}000$$
$$125y=11{,}500$$
$y=92$ lb @ \$2.20
$x=8$ lb @ \$.95

12.

$$x+y=40{,}000$$
$$3y-2{,}000=x$$

$$x+y=40{,}000$$
$$-x+3y=2{,}000$$
$$4y=\$42{,}000$$
$y=\$10{,}500$ son's share
$x=\$29{,}500$ mother's share

13. (Multiply by -8)

$$x+y=100{,}000$$
$$8x+12y=1{,}040{,}000$$

$$-8x-8y=-800{,}000$$
$$8x+12y=1{,}040{,}000$$
$$4y=240{,}000$$
$y=\$60{,}000$ invested in 12% AA Bonds
$x=\$40{,}000$ invested in 8% preferred stock

14. (Multiply by -5)

$$1x+8y=911$$
$$5x+5y=1{,}440$$

$$-5x-40y=-4{,}555$$
$$5x+5y=1{,}440$$
$$-35y=-3{,}115$$
$y=\$.89$ sugar per lb
$x=\$1.99$ coffee per lb

15.

Let x = pizzas
Let y = colas

$$3x+6y=\$13.96$$
$$2x+5y=\$\ 9.65$$

(Multiply by −2) $-6x-12y=-\$27.90$
(Multiply by 3) $6x+15y=\ \$28.95$

$$3y=\ \$\ 1.05$$
$y=.35$ cost of 1 cola

$$2x+5y=9.65$$
$$2x+5(.35)=9.65$$
$$2x+1.75=9.65$$
$$2x=9.65-1.75$$
$$2x=7.90$$
$x=3.95$ cost of 1 pizza

16. (Multiply by −7) $x+y=14{,}800$

$$7x+6\tfrac{1}{2}y=100{,}200$$

$$-7x-7y=-103{,}600$$
$$7x+6\tfrac{1}{2}y=100{,}200$$

$$-\tfrac{1}{2}y=-3{,}400$$
$y=\$6{,}800$ invested in preferred stocks
$x=\$8{,}000$ invested in Treasury bonds

17. (Multiply by −2) $6x+12y=11.10$

$$12x+6y=10.86$$

$$-12x-24y=-22.20$$
$$12x+6y=10.86$$

$$-18y=-11.34$$
$y=\$.63$ cost of 1 can of tomatoes
$$6x+7.56=11.10$$
$$6x=3.54$$
$x=\$.59$ cost of 1 can of peaches

18. (Multiply by 3) $10x+8y=970$
(Multiply by −4) $12x+6y=1{,}020$

$$30x+24y=2{,}910$$
$$-48x-24y=-4{,}080$$

$$-18x=-1{,}170$$
$x=\$65$ men's daily wage
$$650+8y=970$$
$$8y=970-650$$
$$8y=320$$
$y=\$40$ boy's daily wage

19. (Multiply by −20) $x+y=\$46{,}000$

$$20x+8y=\$560{,}000$$

$$-20x-20y=-920{,}000$$
$$20x+8y=560{,}000$$

$$-12y=-360{,}000$$
$y=\$30{,}000$ @ 8%
$x=\$16{,}000$ @ 20%

20. (Multiply by -7)

$$x+y=30{,}000$$
$$5\tfrac{1}{2}x+7y=183{,}000$$

$$-7x-7y=-210{,}000$$
$$\underline{5\tfrac{1}{2}x+7y=183{,}000}$$
$$-1\tfrac{1}{2}x=-27{,}000$$
$$-\tfrac{3}{2}x=-27{,}000$$
$$x=\$18{,}000 \text{ @ } 5\tfrac{1}{2}\%$$
$$y=\$12{,}000 \text{ @ } 7\%$$

21. Equation (1) $x+y=10$ ft
Equation (2) $40x=10y$

(Multiply by -40) (1) $x+y=10$ ft
(2) $40x-10y=0$

(1) $-40x-40y=-400$
(2) $\underline{40x-10y=0}$

$$-50y=-400$$
$y=8$ ft distance from fulcrum
$x=2$ ft distance from fulcrum

22. (Multiply by -150)

$$x+y=6{,}000$$
$$1.50x+3.00y=14{,}250.00$$

$$-150x-150y=-900{,}000$$
$$\underline{150x+300y=1{,}425{,}000}$$
$$150y=525{,}000$$
$1y=3{,}500$ adults
$1x=2{,}500$ children

23. Let $x=$ rate of faster car
Let $y=$ rate of slower car

(Multiply by -5) $x-y=15$ km
$5x+5y=575$ km

$-5x+5y=-75$
$\underline{5x+5y=575 \text{ km}}$
$10y=500$
$y=50$ km rate of slower car per hour
$x=65$ km rate of faster car per hour

24. Let $x=$ number of adults attending
Let $y=$ number of "others" attending

(Multiply by -5) $x+y=3{,}702$
$5x+2y=16{,}404$

$$
\begin{aligned}
-5x-5y&=-18{,}510\\
5x+2y&=-16{,}404\\
\hline
-3y&=-2{,}106\\
y&=702 \text{ "others"}\\
x&=3{,}000 \text{ adults}
\end{aligned}
$$

25. Let $x =$ speed in still air
Let $y =$ wind velocity

$$
\begin{aligned}
x+y&=770\\
x-y&=680\\
\hline
2x&=1{,}450\\
x&=725 \text{ mph in still air}\\
y&=45 \text{ mph wind velocity}
\end{aligned}
$$

26. Let $x =$ weight of first football player
Let $y =$ weight of second football player

$$
\begin{aligned}
x+y&=396\\
6x&=5y
\end{aligned}
$$

(Multiply by 5)

$$
\begin{aligned}
x+y&=396\\
6x-5y&=0
\end{aligned}
$$

$$
\begin{aligned}
5x+5y&=1{,}980\\
6x-5y&=0\\
\hline
11x&=1{,}980\\
x&=180 \text{ lbs of one player}\\
y&=216 \text{ lbs of other player}
\end{aligned}
$$

27. Let $x =$ number of Liberal Arts pledges
Let $y =$ number of Engineering School pledges

(Multiply by -100)

$$
\begin{aligned}
x+y&=4{,}200\\
150x+100y&=545{,}000
\end{aligned}
$$

$$
\begin{aligned}
-100x-100y&=-420{,}000\\
150x+100y&=545{,}000\\
\hline
50x&=125{,}000\\
x&=2{,}500 \text{ Liberal Arts pledges}\\
y&=1{,}700 \text{ Engineering School pledges}
\end{aligned}
$$

28. Let $x =$ higher test score
Let $y =$ lower test score

$$\frac{x+y}{2}=83.5$$

$$
\begin{aligned}
x-y&=5\\
x+y&=167\\
x-y&=5\\
\hline
2x&=172
\end{aligned}
$$

$x = 86$ higher test score
$y = 81$ lower test score

29.

$$x + y = 36$$
$$\frac{x+4}{2} = \frac{2y}{2} - 1$$

$$x + y = 36$$
$$x + 4 = 2y - 2$$

$$x + y = 36$$

(Multiply by −1) $x - 2y = -6$

$$x + y = 36$$
$$-x + 2y = +6$$

$$3y = 42$$

$y = 14$ Rosalie's age
$x = 22$ Ray's age

30.

$$\frac{x+5}{5} = \frac{y}{4} - \frac{1}{1}$$

$$\frac{2}{5}(y+4) = \frac{2x}{5} + 2$$

(Multiply by 2) $-2x + 2y = 2$

$$4x - 5y = -40$$
$$-4x + 4y = +4$$

$$-y = -36$$

$y = 36$ Larry's age
$x = 35$ Lorna's age

16 Word Problems Review Test

Use one or two variables to solve each of the following problems. Check your answers with the solutions on pages 124–146.

1. The length of a rectangle is 10 less than 3 times the width. If the perimeter is 140 cm, what are the dimensions of the rectangle?

2. Angle A is 3 times angle B and angle B is twice angle C. How many degrees are there in each angle of triangle ABC? (There are 180 degrees in every triangle.)

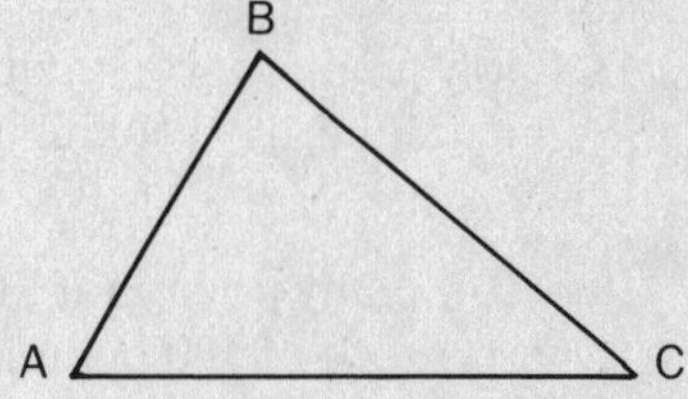

3. Angle E is 10 degrees larger than angle F. Angle D is 3 times angle F less 20 degrees. How many degrees are there in each angle of triangle DEF?

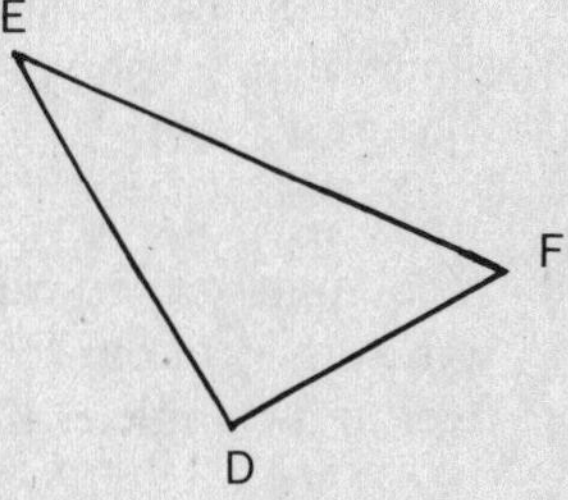

4. The length of a rectangle is 10 less than the width. If the perimeter is 140 centimeters, what are the dimensions of the rectangle?

5. The perimeter of an equilateral triangle (3 equal sides) is 50 inches longer than the length of one side. Find the length of the side.

6. John has 98 coins in his bank made up of dimes and quarters. The total value of the coins is $18.50. How many of each coin does he have?

7. A man has 51 coins made up entirely of nickels and half-dollars. Their value is $14.25. How many of each coin does he have?

8. Divide 5 dollars into dimes and quarters so that there will be 8 more dimes than quarters.

9. John is 6 years older than James. In 5 years John will be 3 times as old as James was 3 years ago. What are their present ages?

10. A father is now 7 times as old as his son. In 5 years he will be 4 times as old as his son will be then. What are their present ages?

11. Frank is 4 years older than John. Five times Frank's age 2 years from now will equal 7 times John's age 2 years ago. What are their present ages?

12. In 4 years, Nancy will be as old as Jane is now. Seven years ago the sum of their ages was 18. Find their ages.

13. A woman has 2 investments which yield an annual income of $996.00. One investment yields 7% and the other 8%. How much is invested in each if the total amount invested is $12,800?

14. Mr. Rosenberg invested part of his $40,000 investment in high yield bonds paying 9½% and preferred stock paying 8%. If his annual income from both investments is $3,620.00, how much did he invest in the bonds and in the stocks?

15. How many ounces of an alloy containing 35% aluminum must be melted with an alloy containing 65% aluminum to obtain 30 ounces of an alloy containing 55% aluminum?

16. How many pounds of $1.60 tea must be added to tea costing $1.80 per pound to make a mixture of 100 pounds costing $1.75 per pound?

17. Mr. Washburn, a plumber, can complete a job in 6 hours and his helper can do the same job in 9 hours. If they both work together to complete the job, how many hours will it take them?

18. A computer expert can do a job on the computer in 6 hours. His secretary can do the same job in 7 hours. After the expert works on the job for 2 hours, his secretary takes over and completes the job. How many hours did it take her to complete the job?

19. How many quarts of pure water must be added to 15 quarts of a solution that is 15% acid in order to dilute it to a solution that is only 12% acid?

20. How much water must be evaporated from a 5% salt solution weighing 60 pounds to obtain a 10% salt solution?

21. Two pipes can fill a tank in 10 hours and 12 hours respectively. A third pipe can drain the tank in 15 hours. How long will it take the tank to be filled if all three pipes are open?

22. An experienced typist can type a legal brief in 6 hours. Her assistant can type the same brief in 9 hours. If they both work together for 3 hours and then the experienced typist works alone, how many hours will it take her to finish the job?

23. Two routes to get from Ventura, California to San Diego, California differ by 20 miles. The time required is the same if a car travels 50 miles per hour over the shorter route and 55 miles per hour over the longer route. How long is each route in miles?

24. Two motorcyclists start toward each other from towns 318 miles apart. If one motorcyclist travels 52 miles per hour and the other 54 miles per hour, in how many hours will they meet?

25. It takes Jake twice as long to harvest a cornfield as it does Arthur. If Arthur and Jake work together, they can harvest the field in 3 days. How many days would it take each alone to harvest the crop?

26. If a weight of 130 pounds is 14 inches from the fulcrum, how far must a weight of 140 pounds be placed from the fulcrum to maintain balance?

27. Two tuna boats leave the same port traveling in opposite directions along the west coast of the United States. One boat travels 5 miles per hour faster than the other. At the end of one day's travel they are 1,080 miles apart. What is the speed of each boat per hour?

28. The charge for admission to the natural history museum is $1.50 for adults and 1 dollar for children. If 5,000 people visited the museum in one day and the total receipts were $6,500, how many of each kind of ticket was sold?

29. A college student jogs from his home to the beach at 8 miles per hour. He visits with his friends on the beach for 4 hours, and then his friends drive him home at the rate of 40 miles per hour. If the student returns home after 6 hours, what is the distance from his home to the beach?

30. A financial consultant invests $6,000 at a certain rate and $8,000 at a rate that is 1% less than twice the rate at $6,000. If the total income from both investments is $1,240.00, find the rate of each investment.

31. A canoeist paddled down a river a distance of 2 miles in 45 minutes. Paddling upstream on his return, it took him 90 minutes. Find the rate of the canoe in still water.

32. A mason contractor needed to build a brick warehouse in 2 days. The first day he hired 8 bricklayers and 4 apprentices at a cost of $1,280 per day. The second day he hired 10 bricklayers and 6 apprentices at a cost of $1,680 per day. What did it cost him per day to pay each bricklayer and each apprentice?

33. Mr. Randolph, a businessman, flew his jet plane from Newport Beach, California to San Francisco, a distance of 480 miles, in 1½ hours against the wind. On his return trip, he covered the same distance in 1 hour. If the speed of the wind was the same going and returning, what was the speed of the tailwind?

34. The perimeter of a triangle is 173 centimeters. The first side is 10 inches greater than the second side. The third side is 5 inches less than twice the first side. Find the length of each side of the triangle.

35. When each side of a square is decreased by 10 centimeters, the area is decreased 300 centimeters^2. Find the length of the original side of the square.

36. Suzanne is one-sixth as old as her mother. In 10 years she will be three-eighths as old as her mother will be then. What are their present ages?

37. A plumber can do a job in 4 hours. When he and his 2 helpers do the job, they can finish it in 2 hours. If both helpers work at the same rate, how many hours would each helper take to do the job alone?

38. Two giant wrestlers weighed a total of 450 pounds. To make a certain weight division, they both decided to go on a diet. One wrestler gained 15% of his weight and the other lost 10% of his weight. If the sum of their weights is now 461.25 pounds, how many pounds did each weigh before starting his diet?

39. Some oil companies are now using power recovery turbines that capture vast amounts of energy that previously went up the stack as waste gas. This device would save approximately 42 million gallons of gas per year per company. If the number of the present oil companies were increased by one-fifth less 2, the number of oil companies would total 22.

a) How many oil companies are there presently?
b) How many gallons of gas will be saved if the new companies use this energy saving device?
c) How many gallons of gasoline would be saved with just the present number of oil companies using this device?

40. Textile workers in the South succeeded in their drive to increase the number of new union members to 6,000 less than one-fifth of their former membership. If their total membership now is 60,000, how many new members did their drive produce? How many were enrolled in their former membership?

41. A seed store has 300 kilograms of a lawn seed mixture worth $2.50 per kilogram. If the mixture was made up of bluegrass selling at $3.00 per kilogram and rye grass selling for $1.75 per kilogram, how many kilograms of each kind of grass were in the mixture?

42. A dairyman has 125 gallons of milk testing 6% butterfat. How many gallons of skimmed milk containing no butterfat must be added to reduce the butterfat content to 5%?

43. A grocer wishes to combine one grade of coffee worth $2.10 per pound with another grade worth $1.90 per pound to obtain 300 pounds worth $2.00 per pound. How many pounds of each grade did he mix?

44. A businessman invested part of $40,000 in stocks paying 6% and the remainder in bonds paying 7%. If his yearly income from both investments is $2,720.00, how much did he invest in stocks and in bonds?

45. A man invested $1,500 more at 6% than at 4%. If his yearly income from both investments is $590.00, how much did he invest at each rate?

46. Two sums of money differing by $1,500 were invested, the larger amount at 5% and the smaller at 6%. If the difference in their yield is $46.00, how much was invested at each rate?

47. A total of $40,000 is invested, part at 4½% and the other part at 6%. The income from both investments is $2,130.00. How much was invested at each rate?

48. Two cyclists leave the same place going in the same direction. One cyclist travels 30 miles per hour and the second travels 35 miles per hour. If the second cyclist starts 2 hours later than the first, in how many hours will the second overtake the first?

49. Two motorbikes leave Atlanta, Georgia at the same time, one traveling north at the speed of 50 miles per hour and the other south at the rate of 55 miles per hour. After how many hours will they be 420 miles apart?

50. A vintage car leaves Des Moines, Iowa traveling at the rate of 45 miles per hour. One hour later a faster car sets out from the same town to overtake the first. At what rate must the faster car travel in order to overtake the first in 5 hours?

51. Two speed boats leave the port of San Francisco traveling in opposite directions at the same time. One boat travels 8 knots per hour faster than the other. After one day's travel they are 1,920 nautical miles apart. What are the speeds of the 2 boats?

52. In a walkathon, two men 27.5 miles apart start walking toward each other. One travels 10 miles in 4 hours and the other travels at the rate of 6 miles in 2 hours. How many miles will each have traveled when they meet?

53. A rocket travels twice the speed of a supersonic plane. If at the end of 1 hour they are 4,680 miles apart, find the speed of each if both flew in opposite directions from the same air field.

54. A mason, working alone, can tile a bathroom in 3 days. He has a helper who can tile the same job alone in 6 days. If the mason and his helper work together, how many days will it take them to do the job?

55. It takes Mary 6 hours to do all the house cleaning that Jane can do in 4 hours and Sue can do in 8 hours. If all 3 work together, how many hours will it take them?

56. John can mow a front and back lawn in 3 hours. When his brother Bill helps him, together they can complete the mowing in 2 hours. How long will it take Bill to do the mowing alone?

57. An oil tank can be filled with oil in 8 hours by one pipe. Another pipe can fill it in 10 hours. If both pipes are used to fill the tank, how many hours will it take them?

58. One large pipe can fill a swimming pool in 10 hours and another pipe can drain the pool in 15 hours. If both pipes are opened, how many hours will it take before the pool is filled?

59. Ralph can paint a house in 4 days. His brother Walter can paint the same house in 6 days. After working 1 day, Ralph became ill and Walter finished the job. How long did it take Walter to finish the job?

60. A storage pipe can fill a reservoir in 15 days. Another smaller pipe can fill it in 20 days. After the larger pipe is on for 2 days, it breaks down. The smaller pipe is opened and completes the job of filling the reservoir. How long did it take for the smaller pipe to complete the job alone?

61. In 1980 the price of gold soared to a new high of $900 per ounce. This is 4½ times what it was in 1972. What was the price per ounce in 1972?

Hint: Let $x =$ price per ounce in 1972

62. A Houston-based firm experimenting with gene splicing needs to increase its office space one and one-half times its present space. If they need 500,000 square feet of space, how many square feet do they have at the present and by how many square feet must they increase their space?

Let $x =$ number of square feet at the present
? = number of square feet to be added

63. In 1982, the American car manufacturers suffered a decline in car sales that amounted to ⅙ of the cars sold in 1979 less 3,000. If the number of cars sold in 1979 less the decline in car sales in 1982 amounted to 300,000 units, what was the decline in car sales in 1982? What were the car sales in 1979?

64. A private environmental protection agency made a study which ascertained that the air quality would be substantially improved in Los Angeles if there were 10,000 less than ⅕ of the present number of cars on the roads. If the present number of cars on the roads less those which should not be on the roads is 480,000 units, how many cars are on the roads and by what amount should they be reduced?

Hint: Let $x =$ number of cars present on the roads
? = number of cars that should be reduced

65. A tennis court has a perimeter of 212 feet. If the length is 24 feet longer than the width, what are the dimensions of the court? (Draw a rectangle.)

66. The radius of the larger circle $\overline{OB}$ is 14 feet and the radius of the smaller circle $\overline{OC}$ is 7 feet. What is the area of the shaded part? (Use $\pi = 22/7$. Area of the circle $= \pi r^2$.)

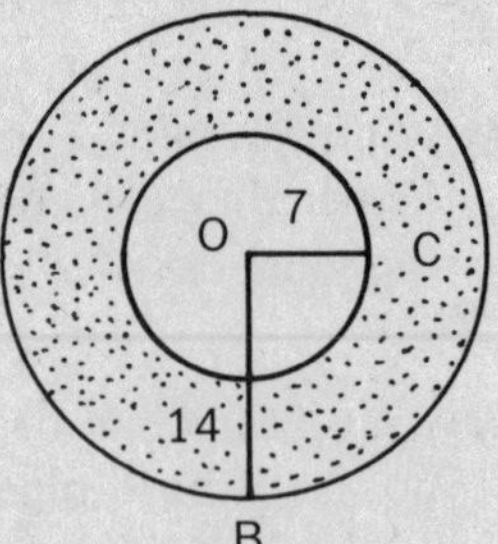

67. A navy boat goes 50 miles downstream in 2½ hours. Going upstream it traveled the same distance in 4 hours. What is the rate of the boat in still water, and what is the rate of the current? (Draw model.)

68. Jess has twice as many dimes as quarters. If the dimes were quarters and the quarters were dimes, he would have $1.05 more than he has now. How many of each kind of coin does he have?

Let x = quarters
$2x$ = dimes
$25x$ = value in cents of quarters
$20x$ = value in cents of dimes

$45x$ = total value of coins

Reverse order:

Let x = dimes
$2x$ = quarters
$10x$ = value of dimes in cents
$50x$ = value of quarters in cents

$60x$ = total value of coins

69. A Federal Railways bus leaves Cleveland for Springfield and averages 56 miles per hour. Thirty minutes later a second bus leaves from the same station for Springfield averaging 60 miles per hour. How long will it take the second bus to overtake the Federal Railways bus?

70. An attendant at a service station checks a truck radiator and finds it to contain 40% antifreeze. If the radiator holds 20 quarts and is full, how many quarts must be drained off and replaced in order to get a 50% antifreeze reading?

71. The sum of the ages of two men is 62 years. One-half of the younger's age 6 years from now is equal to one-third of the older's age 2 years from now. How old is each man now?

Let x = age of older man
Let y = age of younger man

72. How many degrees in each angle of the triangle?

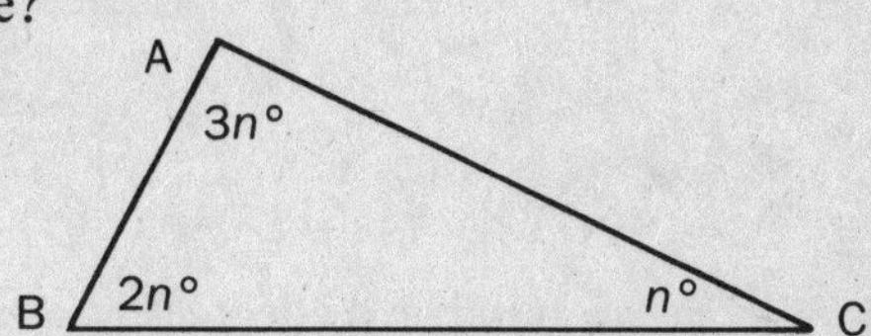

73. AOB is the diameter of the circle. Using $C = \pi d$ and $\pi = 22/7$, find the length of the diameter in centimeters if the circumference is 154 centimeters. (Use $C = \pi d$.)

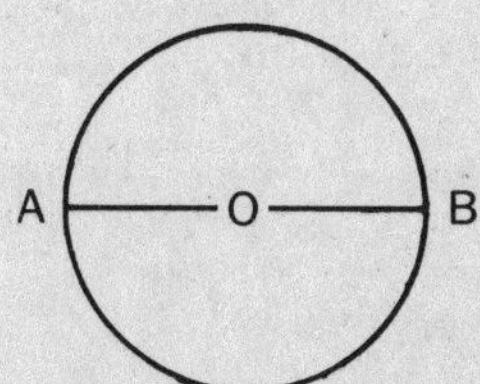

74. A worn-out pump can empty a tank in 8 hours. A new pump can empty the tank in 4 hours. After the old pump is used for 3 hours, the new pump replaces it and empties the tank. How long did it take for the new pump to empty the tank? (Draw model.)

75. A bricklayer can build a wall in 8 days. His helper can build the same wall in 10 days. After the bricklayer works for 3 days, his helper takes over and completes the job. How long did it take him? (Draw model.)

76. A man 6 feet tall casts a shadow 14 feet. At the same time, a telephone pole 60 feet tall would cast a shadow __?__ feet.

77. The ratio of two numbers is 7 : 4. If the difference of the 2 numbers is 45, what are the numbers?

78. Mr. Santos is 28 years older than his son Roberto. In 12 years Mr. Santos will be twice as old as Roberto will be then. How old is each now?

79. A jet travels 500 kilometers in 40 minutes with a tail wind. Returning, the jet takes 50 minutes to cover the same distance. What is the rate of the plane and the speed of the wind?

80. If 40 men working on a U.S. Government project can complete the job in 100 hours, how many men would be required to complete the job in 80 hours?

Answers to Word Problems—Review Test

For additional help, turn to the problems indicated in the box to the right of each solution.

1. Let $x =$ width of the rectangle
Let $3x - 10 =$ length of the rectangle

Refer to Problem 2 Chapter 3

Equation:
$$2x + 2(3x - 10) = 140 \text{ cm}$$
$$2x + 6x - 20 = 140$$
$$8x - 20 = 140$$
$$8x = 160$$
$$x = 20 \text{ cm in width}$$
$$(3x - 10) = 50 \text{ cm in length}$$

2. Let $x =$ number of degrees in angle C
$2x =$ number of degrees in angle B
$6x =$ number of degrees in angle A

Refer to Problem 4 Chapter 3

Equation:
$$9x = 180°$$
$$x = 20° \text{ in angle C}$$
$$2x = 40° \text{ in angle B}$$
$$6x = 120° \text{ in angle A}$$

3. Let $x =$ degrees in angle F
$(x + 10) =$ degrees in angle E
$(3x - 20) =$ degrees in angle D

Refer to Problem 4 Chapter 3

Equation:
$$x + x + 10 + 3x - 20 = 180$$
$$5x - 10 = 180$$
$$5x = 190$$
$$x = 38° \text{ in angle F}$$
$$x + 10 = 48° \text{ in angle E}$$
$$3x - 20 = 94° \text{ in angle D}$$

4. Let $x =$ number of cm in width
$x - 10 =$ number of cm in length

Refer to Problem 3 Chapter 3

Equation:

$$\begin{aligned} 2x + 2(x - 10) &= 140 \text{ cm} \\ 2x + 2x - 20 &= 140 \text{ cm} \\ 4x &= 140 + 20 \\ 4x &= 160 \text{ cm} \\ x &= 40 \text{ cm width} \\ x - 10 &= 30 \text{ cm length} \end{aligned}$$

5. Let $s =$ length of side of equilateral triangle in inches
$3s =$ perimeter of the triangle in inches

Refer to Problem 2 Chapter 3

Equation:

$$\begin{aligned} 3s &= 50 \text{ in.} + s \\ 3s - s &= 50 \text{ in.} \\ 2s &= 50 \text{ in.} \\ s &= 25 \text{ in. length of side of equilateral triangle} \end{aligned}$$

6. Let $x =$ number of dimes
$98 - x =$ number of quarters
$10x =$ value in cents of dimes
$25(98 - x) =$ value in cents of quarters
$\$18.50 = 1{,}850$ value in cents

Refer to Study Problem 1 Chapter 5

Equation:

$$\begin{aligned} 10x + 25(98 - x) &= 1{,}850 \\ 10x + 2{,}450 - 25x &= 1{,}850 \\ -15x &= 1{,}850 - 2{,}450 \\ -15x &= -600 \\ x &= 40 \text{ dimes} \\ 98 - x &= 58 \text{ quarters} \end{aligned}$$

7. Let $x =$ number of nickels
$y =$ number of half-dollars

Refer to Study Problem 11 Chapter 15

(Multiply by -5)

$$\begin{aligned} x + y &= 51 \\ 5x + 50y &= 1{,}425 \text{ value in cents} \\ -5x - 5y &= -255 \\ 5x + 50y &= 1{,}425 \\ \hline 45y &= 1{,}170 \\ y &= 26 \text{ half-dollars} \\ x &= 25 \text{ nickels} \end{aligned}$$

8. Let $x =$ number of quarters
Let $x + 8 =$ number of dimes
$25x =$ value of quarters in cents
$10(x + 8) =$ value of dimes in cents
$\$5.00 = 500$ value in cents

Refer to Study Problem 1 Chapter 5

Equation:

$$\begin{aligned} 25x + 10(x + 8) &= 500 \\ 25x + 10x + 80 &= 500 \\ 35x &= 500 - 80 \\ 35x &= 420 \\ x &= 12 \text{ quarters} \\ x + 8 &= 20 \text{ dimes} \end{aligned}$$

9.

	Age now	Age in 5 yr	Age 3 yr ago
John	$x+6$	$x+11$	
James	x		$x-3$

Equation:

$$x+11=3(x-3)$$
$$x+11=3x-9$$
$$-2x=-20$$
$$x=10 \text{ James's age}$$
$$x+6=16 \text{ John's age}$$

Refer to Study Problem 4 Chapter 11

10.

	Age now	Age in 5 yr
Father	$7x$	$7x+5$
Son	x	$x+5$

Equation:

$$7x+5=4(x+5)$$
$$7x+5=4x+20$$
$$3x=15$$
$$x=5 \text{ son's age}$$
$$7x=35 \text{ father's age}$$

Refer to Study Problem 3 Chapter 11

11.

	Age now	Age 2 yr from now	Age 2 yr ago
Frank	$x+4$	$x+6$	
John	x		$x-2$

Equation:

$$5(x+6)=7(x-2)$$
$$5x+30=7x-14$$
$$5x-7x=-14-30$$
$$-2x=-44$$
$$x=22 \text{ John's age now}$$
$$x+4=26 \text{ Frank's age now}$$

Refer to Study Problem 4 Chapter 11

12.

	Age now	Age 7 yr ago
Nancy	$x-4$	$x-11$
Jane	x	$x-7$

Equation:

$$x-11+x-7=18$$
$$2x-18=18$$
$$2x=36$$
$$x=18 \text{ Jane's age now}$$
$$x-4=14 \text{ Nancy's age now}$$

Refer to Study Problem 3 Chapter 11

13.

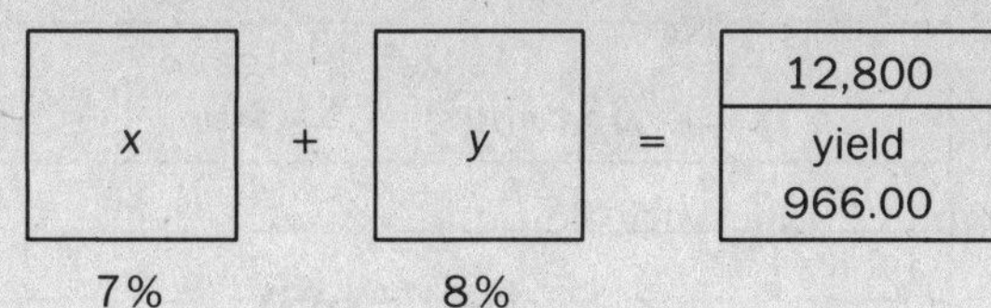

Refer to Study Problem 12 Chapter 15

Equations: (1) $x+y=12{,}800$

(2) $.07x+.08y=966.00$

Move decimal point 2 places to the right.

Multiply (1) by -7 $\quad -7x-7y=-89{,}600$

$\quad 7x+8y=96{,}600$

Add (1) and (2) $\quad 1y=\$7{,}000$ invested @ 8%

Substitute \$7,000 for y in equation (1)

$x+7{,}000=12{,}800$

$x=\$5{,}800$ invested @ 7%

14.

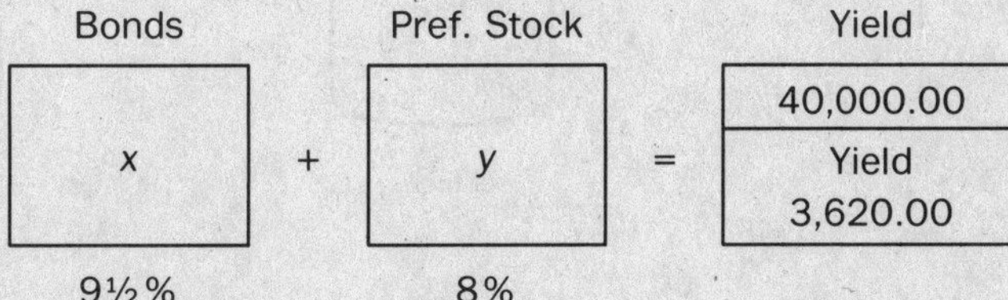

Refer to Study Problem 12 Chapter 15

Equations: (1) $x+y=40{,}000$

(2) $.09\frac{1}{2}x+.08y=3620.00$

Move decimal point 2 places to the right.

$9\frac{1}{2}x+8y=362{,}000$

Multiply (1) by -8

$-8x-8y=-320{,}000$

$9\frac{1}{2}x+8y=362{,}000$

Add (1) and (2) $\quad 1\frac{1}{2}x=42{,}000$

$\frac{3}{2}x=42{,}000$

$x=\$28{,}000$ invested in 9½% bonds

Substitute \$28,000 for x in (1)

$28{,}000+y=40{,}000$

$y=\$12{,}000$ invested in 8% preferred stock

15.

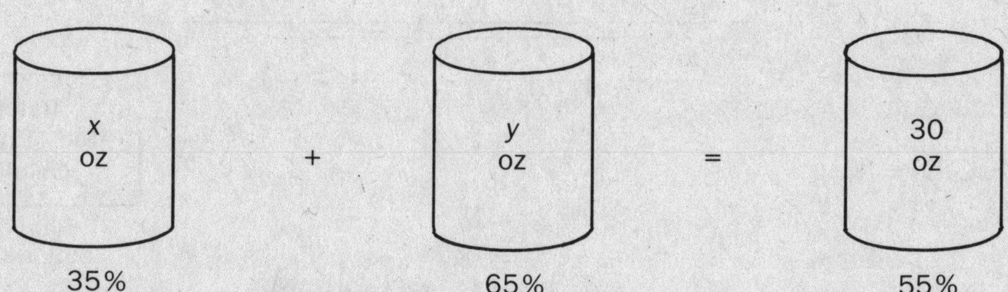

Equations: (1) $x+y=30$

(2) $.35x+.65y=.55(30)$

Refer to Study Problem 10 Chapter 15

Move decimal point 2 places to the right.

$$35x+65y=1{,}650$$

Multiply (1) by -35

$$-35x-35y=-1{,}050$$
$$35x+65y=1{,}650$$

Add (1) and (2)

$$30y=600$$
$$y=20 \text{ oz of } 65\%$$

Substitute 20 for y in (1)

$$x+20=30$$
$$x=10 \text{ oz of } 35\%$$

16.

$1.60

+

$1.80

=

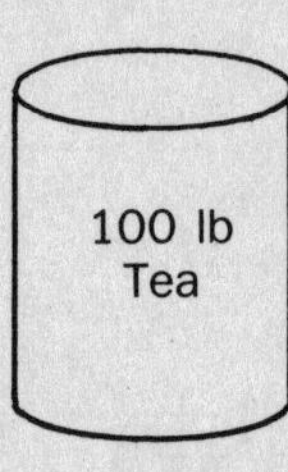

$1.75

Equations: (1) $x+y=100$

(2) $1.60x+1.80y=1.75(100)$

Refer to Study Problem 9 Chapter 15

Move decimal point 2 places to the right.

$$160x+180y=17{,}500$$

Multiply (1) by -160

$$-160x-160y=-16{,}000$$
$$160x+180y=17{,}500$$

Add (1) and (2)

$$20y=1{,}500$$
$$y=75 \text{ lb of } \$1.80 \text{ tea}$$

Substitute 75 for y in (1)

$$x+75=100$$
$$x=25 \text{ lb of } \$1.60 \text{ tea}$$

17.

	Time to do job	*Part done in 1 hr*
Plumber	6 hr	1/6
Helper	9 hr	1/9
Together	x hr	$1/x$

Equation:

$$\frac{1}{6}+\frac{1}{9}=\frac{1}{x}$$

Refer to Study Problem 1 Chapter 8

L.C.D. $=18x$

$$3x+2x=18$$
$$5x=18$$
$$x=3\tfrac{3}{5} \text{ hr together}$$

18.

	Hours to finish job	*Part done in 1 hr*
Expert	6	1/6
Secretary	7	1/7
Together	x	$1/x$
Part of job completed 1/3 by expert		

Equation: $\frac{1}{7} = \frac{2}{3x}$

L.C.D. $= 21x$

Refer to Study Problem 1 Chapter 8

$3x = 14$

$x = 4\frac{2}{3}$ hr for secretary to complete job

19.

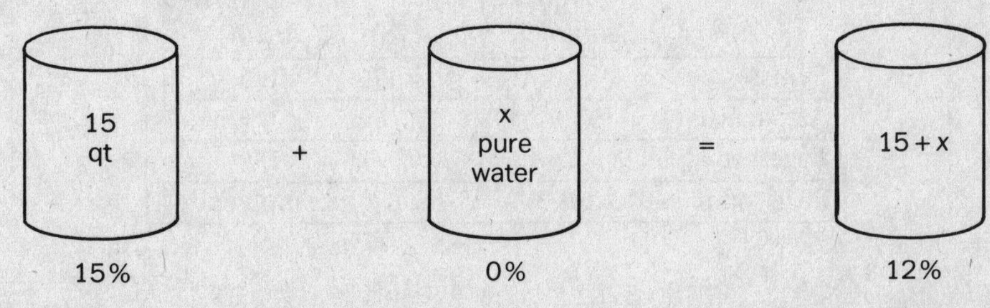

$.15 \qquad\qquad .12$

Move decimal point 2 places to the right.

Refer to Study Problem 3 Chapter 6

$225 + 0x = 180 + 12x$

$45 = 12x$

$x = 3\frac{3}{4}$ quarts

20.

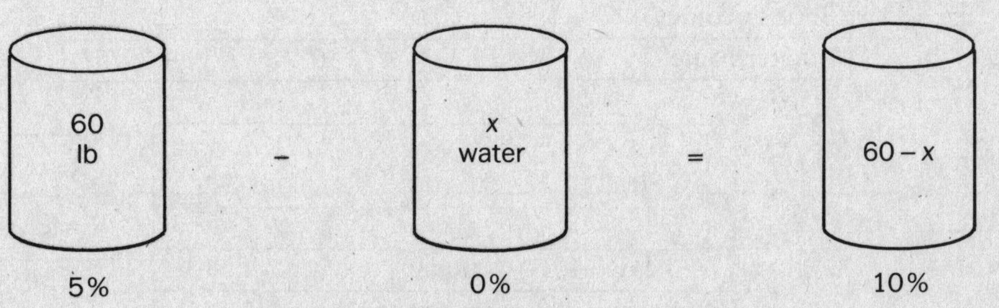

$.05(60) = .10(60 - x)$

Move decimal point 2 places to the right.

Refer to Study Problem 3 Chapter 6

$300 = 600 - 10x$

$10x = 300$

$x = 30$ lb to be evaporated

21.

	Hours to fill tank	*Part filled in 1 hr*
1st pipe	10 hr	1/10
2nd pipe	12 hr	1/12
3rd pipe	15 hr (drain)	−1/15 drained
Together	x	$1/x$

Equation: $$\frac{1}{10}+\frac{1}{12}+\frac{-1}{15}=\frac{1}{x}$$

L.C.D. $=60x$

Refer to Study Problem 3 Chapter 8

$$6x+5x-4x=60$$
$$7x=60$$
$$x=8\tfrac{4}{7} \text{ hr to fill tank}$$

22.

	Hours to complete job	*Part completed in 1 hr*
Typist	6 hr	1/6
Assistant	9 hr	1/9
Together	x hr	$1/x$
Part completed $3/6+3/9=5/6$ of job completed		

Equation: $$\frac{1}{6}=\frac{1}{6x}$$

L.C.D. $=6x$

Refer to Study Problem 3 Chapter 8

$$\frac{\not{6}x(1)}{\not{6}}=\frac{\not{6}\not{x}(1)}{\not{6}\not{x}}$$

$x=1$ hr for experienced typist to finish job

23.

	Distance	*Rate*	*Time*
Shorter route	x	50	$x/50$
Longer route	$x+20$	55	$x+20/55$

Equation: $$\frac{x}{50}=\frac{x+20}{55}$$

L.C.D. $=550$

Refer to Study Problem 2 Chapter 4

$$11x=10x+200$$
$$x=200 \text{ miles for shorter route}$$
$$x+20=220 \text{ miles for longer route}$$

24.

	Distance	*Rate*	*Time*
1st motorcyclist	$52x$	52	x
2nd motorcyclist	$54x$	54	x

Equation:

$$52x+54x=318$$
$$106x=318$$
$$x=3 \text{ hours}$$

Refer to Study Problem 3 Chapter 4

25. First calculate the amount of time it takes for both to do the job together.

Person	*Rate per day*	*Rate each does in 1 day*
Jake	$2x$	$1/2x$
Arthur	x	$1/x$
Together	3	1/3

Equation: $\frac{1}{x}+\frac{1}{2x}=\frac{1}{3}$

L.C.D. $=6x$

Refer to Study Problem 4 Chapter 8

$$6+3=2x$$
$$9=2x$$
$$x=4\tfrac{1}{2} \text{ days for Arthur working alone}$$
$$2x=9 \text{ days for Jake working alone}$$

26.

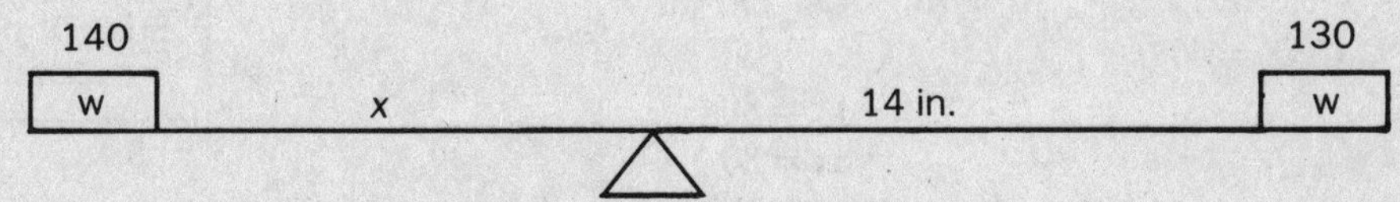

$$140\cdot x=130\cdot 14$$
$$140x=1{,}820$$
$x=13$ inches (distance from fulcrum that 140-pound weight must be placed)

Refer to Study Problem 1 Chapter 10

27.

	Distance	*Rate*	*Time*
Slow tuna boat	$24x$	x	24
Faster tuna boat	$24(x+5)$	$x+5$	24

Equation:

$$24x+24(x+5)=1{,}080 \text{ miles}$$
$$24x+24x+120=1{,}080$$
$$48x+120=1{,}080$$
$$48x=1{,}080-120$$
$$48x=960$$
$x=20$ miles per hour for slow tuna boat

$x+5=25$ miles per hour for faster tuna boat

Refer to Study Problem 1 Chapter 4

28.

Equations: (1) $x+y=5{,}000$

(2) $1.50x+1.00y=6{,}500.00$

Refer to Study Problem 17 Chapter 15

Move decimal point 2 places to the right.

$$150x+100y=650{,}000$$

Multiply (1) by -100 $\quad -100x-100y=-500{,}000$

$$150x+100y=650{,}000$$

Add (1) and (2) $\quad 50x=150{,}000$

$$x=3{,}000 \text{ adults}$$

Substitute 3,000 for x in (1) $\quad 3{,}000+y=5{,}000$

$$y=2{,}000 \text{ children}$$

29.

Jogger	*Distance*	*Rate*	*Time*
To beach	x	8	$x/8$
To home	x	40	$x/40$

Equation: $\frac{x}{8}+\frac{x}{40}=\frac{2}{1}$

Refer to Study Problem 2 Chapter 4

L.C.D. = 40

$$5x+1x=80$$
$$6x=80$$
$$x=13\tfrac{1}{3} \text{ miles distance from his home to the beach}$$

30.

6,000 invested	+	8,000 invested	=	yield 1,240.00
x%		$(2x-1)$%		
$.01(x)$		$.01(2x-1)$		$.01$

Move decimal point 2 places to the right.

Refer to Study Problem 3 Chapter 7

Equation: $6{,}000x+8{,}000(2x-1)=124{,}000$

$$6{,}000x+16{,}000x-8{,}000=124{,}000$$
$$22{,}000x=124{,}000+8{,}000$$
$$22{,}000x=132{,}000$$
$$x=6\% \text{ rate paid on } \$6{,}000$$
$$2x-1=11\% \text{ rate paid on } \$8{,}000$$

31. Let $r=$ rate of canoe in still water

$x=$ rate of current

	Distance	*Rate*	*Time*
Downstream	2	$r+x$	3/4
Upstream	2	$r-x$	3/2

$$\text{Time Downstream} = \frac{D}{r+x}$$

$$\text{Time Upstream} = \frac{D}{r-x}$$

Refer to
Study Problem 8
Chapter 15

Equation 1 (Proportion 1)

$$\frac{3}{4} = \frac{2}{r+x}$$

Equation 2 (Proportion 2)

$$\frac{3}{2} = \frac{2}{r-x}$$

(Product of means equals product of extremes)

$$3(r+x) = 8$$
$$3r + 3x = 8$$

$$3(r-x) = 4$$
$$3r - 3x = 4$$

$$3r + 3x = 8$$
$$3r - 3x = 4$$
$$6r = 12$$

$r = 2$ miles per hour (rate of canoe in still water)

32. Let $x =$ cost of 1 bricklayer's work per day
$y =$ cost of 1 apprentice's work per day

Refer to
Study Problem 7
Chapter 15

(Multiply by -3) $8x + 4y = \$1{,}280$

(Multiply by 2) $10x + 6y = \$1{,}680$

$$-24x - 12y = -3{,}840$$
$$20x + 12y = 3{,}360$$
$$-4x = -480$$

$x = \$120$ cost per day for bricklayer
$y = \$80$ cost per day for the apprentice

33.

	Distance	*Rate*	*Time*
Going	480	$r-x$	1½
Returning	480	$r+x$	1

Equation 1 (Proportion 1)

$$\frac{480}{r-x} = \frac{3}{2}$$

Equation 2 (Proportion 2)

$$\frac{480}{r+x} = \frac{1}{1}$$

(Product of means equals product of extremes)

Equation 1 $3r - 3x = 960$
Equation 2 $r + x = 480$

Refer to
Study Problem 8
Chapter 15

(Multiply by 3)

$$3r - 3x = 960$$
$$3r + 3x = 1{,}440$$
$$6r = 2{,}400$$

$r = 400$ mph
$x = 80$ mph speed of tailwind

34. Let x = length of the second side
$x+10$ = length of the first side
$2(x+10)-5$ = length of the third side

Refer to Problem 3 Chapter 2

Equation:
$$x+x+10+2(x+10)-5=173$$
$$x+x+10+2x+20-5=173$$
$$4x+25=173$$
$$4x=173-25$$
$$4x=148$$
$x=37$ in. length of the second side
$x+10=47$ in. length of the first side
$2(x+10)-5=89$ in. length of the third side

35. Let x = length of side of original square
Let x^2 = area of original square
$(x-10)$ = length of new square
$(x-10)^2$ = area of the new square

Refer to Problem 3 Chapter 2

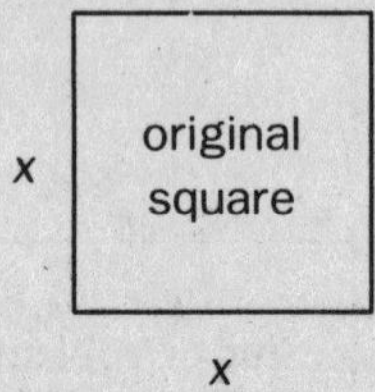

x − 10
new
square
x − 10

Equation:
$$(x-10)^2=x^2-300$$
$$x^2-20x+100=x^2-300$$
$$x^2-x^2-20x=-300-100$$
$$-20x=-400$$
$x=20$ cm length of original square

36.

	Age now	*Age in 10 years*
Suzanne	$1/6x$	$1/6x+10$
Mother	x	$x+10$

Equation:
$$\frac{1}{6x}+10=\frac{3}{8}(x+10)$$

Refer to Study Problem 4 Chapter 11

L.C.D. = 24
$$4x+240=9x+90$$
$$4x-9x=90-240$$
$$-5x=-150$$
$x=30$ years (Mother's age now)
$\frac{1}{6}x=5$ years (Suzanne's age now)

37.

	Hours to do job	*Part done in 1 hr*
Plumber	4 hr	1/4
Helper	x hr	$1/x$
Helper	x hr	$1/x$
Together	2 hr	1/2

Equation: $\frac{1}{4}+\frac{1}{x}+\frac{1}{x}=\frac{1}{2}$

L.C.D. $= 4x$

Clearing fractions $x+4+4=2x$

$8=2x-x$

8 hr $= x$ time for each helper to do the job alone

Refer to Study Problem 1 Chapter 8

38.

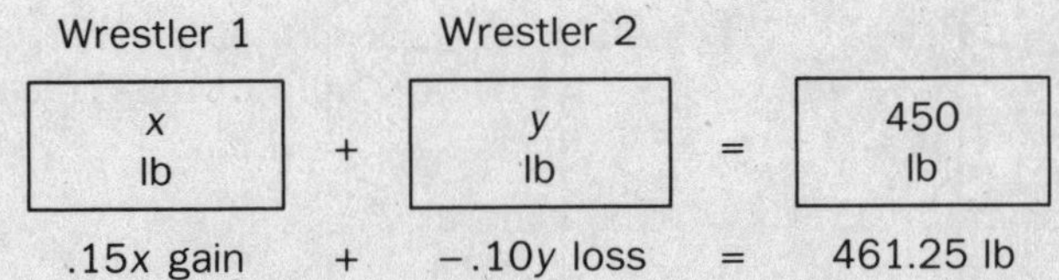

.15x gain + −.10y loss = 461.25 lb

$100\% + 15\%$ gain $= 115\%$

$100\% - 10\%$ loss $= 90\%$

$1.15x + .90y = 461.25$

Refer to Study Problem 14 Chapter 15

Move decimal point 2 places to the right.

$115x+90y=46{,}125$ new weight

(Multiply by -90) $x+y=450$ original weight

$$\begin{array}{r} 115x+90y=46{,}125 \\ -90x-90y=-40{,}500 \\ \hline 25x=5{,}625 \end{array}$$

$x=225$ lb original weight of Wrestler 1

$x+y=450$

$225+y=450$

$y=225$ lb original weight of Wrestler 2

39. Let $x =$ number of oil companies presently
$\frac{1}{5}x-2 =$ increase of oil companies

Refer to Example 1 Chapter 8

Equation: $x+\frac{1}{5}x-2=22$

L.C.D. $= 5$ $5x+1x-10=110$

$6x-10=110$

$6x=120$

$x=20$ oil companies at present

$\frac{1}{5}\cdot 20-2=$ increase

$4-2=2$ increase

a) 20 oil companies presently
b) $2\cdot 42$ million $= 84$ million gallons of gasoline
c) $20\cdot 42$ million equals 840 million gallons of gasoline

40. Let $x =$ former membership
Let $y =$ new membership

Refer to Study Problem 7 Chapter 15

Equations: (1) $x+y=60{,}000$

(2) $y=\frac{1}{5}x-6{,}000$

$$x + y = 60{,}000$$

Multiply (2) by -1: $\quad -\tfrac{1}{5}x + y = -6{,}000$

Add (1) and (2): $\quad 1\tfrac{1}{5}x = 66{,}000$

$x = 55{,}000$ former membership

$y = 5{,}000$ new members

41.

Equations: (1) $x + y = 300$ kg

(2) $3.00x + 1.75y = 750.00$

Refer to Study Problem 10 Chapter 15

Move decimal point 2 places to the right.

Multiply (1) by -300 $\quad -300x - 300y = -90{,}000$

$300x + 175y = 75{,}000$

Add (1) and (2) $\quad -125y = -15{,}000$

$y = 120$ kg of rye grass

$x = 180$ kg of bluegrass

42.

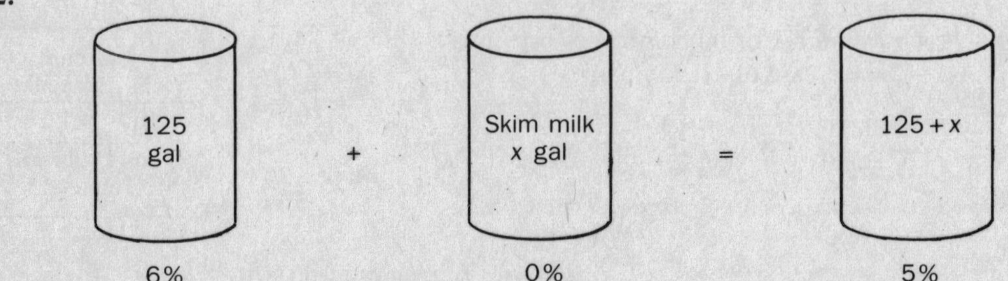

$$.06 + 0 = .05$$

Refer to Study Problem 3 Chapter 6

Move decimal point 2 places to the right.

$$6(125) + 0 = 5(125 + x)$$
$$750 = 625 + 5x$$
$$750 - 625 = 5x$$
$$125 = 5x$$

$25 = x$ gal of skim milk to be added

43.

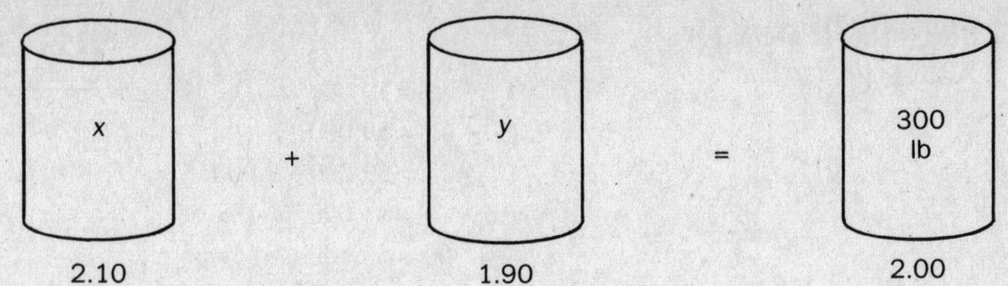

Equations: (1) $x+y=300$ lb

(2) $2.10x+1.90y=2.00(300)$

Refer to Study Problem 7 Chapter 15

Move decimal point 2 places to the right.

$$210x+190y=60{,}000$$

Multiply (1) by -190

$$-190x-190y=57{,}000$$
$$210x+190y=60{,}000$$

Add (1) and (2) $20x=3{,}000$

$x=150$ pounds @ \$2.10 per pound

Substitute 150 for x in (1)

$$150+y=300$$

$y=150$ pounds @ \$1.90 per pound

44.

Stocks x	+	Bonds y	=	40,000 yield 2,720.00
6%		7%		
.06		.07		

Equations: (1) $x+y=\$40{,}000$

(2) $.06x+.07y=\$2{,}720.00$

Refer to Study Problem 12 Chapter 15

Move decimal point 2 places to the right.

$$6x+7y=272{,}000$$

Multiply (1) by -6

$$-6x-6y=-240{,}000$$

Add (1) and (2) $y=\$32{,}000$ invested in bonds at 7%

Substitute 32,000 for y in (1)

$$x+32{,}000=40{,}000$$

$x=\$8{,}000$ invested in stocks at 6%

45.

$x+1{,}500$	+	x	=	590.00
6%		4%		
.06		.04		

$$.06(x+1{,}500)+.04(x)=590.00$$

Move decimal point 2 places to the right.

Refer to Study Problem 3 Chapter 7

$$6(x+1{,}500)+4(x)=59{,}000$$
$$6x+9{,}000+4x=59{,}000$$
$$10x=59{,}000-9{,}000$$
$$x=\$5{,}000 \text{ @ } 4\%$$
$$x+1{,}500=\$6{,}500 \text{ @ } 6\%$$

46.

x	−	y	=	1,500 46.00
5%		6%		
.05		.06		46.00

Move decimal point 2 places to the right.

Refer to Study Problem 17 Chapter 15

Equations: (1) $x-y=1{,}500$

(2) $5x-6y=4{,}600$

Multiply (1) by −5 $-5x+5y=-7{,}500$

$5x-6y=4{,}600$

Add (1) and (2) $-1y=-2{,}900$

$y=\$2{,}900$ invested @ 6%

Substitute 2,900 for y in (1)

$$x-2{,}900=1{,}500$$
$$x=\$4{,}400 \text{ invested @ } 5\%$$

47.

x	+	y	=	40,000 invested 2,130.00 yield
4½%		6%		
.04½		.06		

Move decimal point 2 places to the right.

Refer to Study Problem 10 Chapter 15

Equations: (1) $x+y=\$40{,}000$

(2) $4\frac{1}{2}x+6y=\$213{,}000$

Multiply (1) by −6 $-6x-6y=-240{,}000$

$4\frac{1}{2}x+6y=213{,}000$

Add (1) and (2) $-1\frac{1}{2}x=-27{,}000$

$x=\$18{,}000$ invested @ 4½%

Substitute 18,000 for x in (1)

$$18{,}000+y=40{,}000$$
$$y=\$22{,}000 \text{ invested @ } 6\%$$

48.

	Distance	*Rate*	*Time*
1st cyclist	$30x$	30	x
2nd cyclist	$35(x-2)$	35	$(x-2)$

Equation:

$$30x = 35(x - 2)$$
$$30x = 35x - 70$$
$$30x - 35x = -70$$
$$-5x = -70$$
$$x = 14$$
$x - 2 = 12$ hr for second cyclist to overtake first cyclist

Refer to Study Problem 2 Chapter 4

49.

	Distance	*Rate*	*Time*
1st bike (north)	$50x$	50	x
2nd bike (south)	$55x$	55	x

Equation:

$$50x + 55x = 420$$
$$105x = 420$$
$x = 4$ hours to be 420 miles apart

Refer to Study Problem 1 Chapter 4

50.

	Distance	*Rate*	*Time*
Vintage car	270	45	6
Faster car	$5x$	x	5

Equation:

$$5x = 270$$
$x = 54$ mph for faster car to overtake vintage car

Refer to Study Problem 1 Chapter 4

51.

	Nautical miles	*Knots per hr*	*Hours*
Slow boat	$24x$	x	24
Faster boat	$24(x+8)$	$x+8$	24

Equation:

$$24x + 24(x + 8) = 1{,}920$$
$$24x + 24x + 192 = 1{,}920$$
$$48x = 1{,}920 - 192$$
$$48x = 1{,}728$$
$x = 36$ knots per hr slower boat
$x + 8 = 44$ knots per hr faster boat

Refer to Study Problem 3 Chapter 4

52.

	Distance	*Rate*	*Time*
1st walker	$2\frac{1}{2}x$	$2\frac{1}{2}$	x
2nd walker	$3x$	3	x

Equation:

$$2\tfrac{1}{2}x + 3x = 27.5$$
$$5.5x = 27.5$$
$$55x = 275$$
$x = 5$ hours
$2\frac{1}{2} \cdot 5 = 12\frac{1}{2}$ miles for 1st walker
$3 \cdot 5 = 15$ miles for 2nd walker

Refer to Study Problem 1 Chapter 4

53.

	Miles	*Rate*	*Hours*
Rocket	$2x$	$2x$	1
Plane	x	x	1

Equation:

$$x + 2x = 4{,}680$$
$$3x = 4{,}680$$
$$x = 1{,}560 \text{ mph for plane}$$
$$2x = 3{,}120 \text{ mph for rocket}$$

Refer to
Study Problem 1
Chapter 4

54.

	Days to complete job	*Work done in 1 day*
Mason	3	1/3
Helper	6	1/6
Together	x	$1/x$

Equation: $\frac{1}{3} + \frac{1}{6} = \frac{1}{x}$

L.C.D. $= 6x$

$$2x + x = 6$$
$$x = 2 \text{ days to complete job working together}$$

Refer to
Study Problem 1
Chapter 8

55.

	Hours to complete job	*Work done in 1 hr*
Mary	6	1/6
Sue	8	1/8
Jane	4	1/4
Together	x	$1/x$

Equation: $\frac{1}{6} + \frac{1}{8} + \frac{1}{4} = \frac{1}{x}$

L.C.D. $= 24x$

$$4x + 3x + 6x = 24$$
$$13x = 24$$
$$x = 1\tfrac{11}{13} \text{ hr if all work together to complete the job}$$

Refer to
Study Problem 2
Chapter 8

56.

	Hours to complete job	*Part done in 1 hr*
John	3	1/3
Bill	x	$1/x$
Together	2	1/2

Equation: $\frac{1}{3} + \frac{1}{x} = \frac{1}{2}$

L.C.D. $= 6x$

$$2x + 6 = 3x$$
$$x = 6 \text{ hr for Bill to do the job alone}$$

Refer to
Study Problem 1
Chapter 8

57.

	Hours to fill tank	*Part of tank filled in 1 hr*
One pipe	8	1/8
Other pipe	10	1/10
Together	x	$1/x$

Equation: $$\frac{1}{8}+\frac{1}{10}=\frac{1}{x}$$

L.C.D. $= 40x$

Refer to Study Problem 1 Chapter 8

$$5x+4x=40$$
$$9x=40$$
$$x=4\tfrac{4}{9} \text{ hours to fill tank}$$

58.

	Hours to fill tank	*Part of tank filled in 1 hr*
Large pipe	10	1/10
Drain pipe	15	−1/15
Together	x	$1/x$

Equation: $$\frac{1}{10}-\frac{1}{15}=\frac{1}{x}$$

L.C.D. $= 30x$

Refer to Study Problem 3 Chapter 8

$$3x-2x=30$$
$$x=30 \text{ hours to fill pool}$$

59.

	Days to paint house	*Part painted in 1 day*
Ralph	4	1/4
Walter	6	1/6
Together	x	$1/x$
Part to be completed by Walter		3/4

Equation: $$\frac{1}{6}=\frac{3}{4x}$$

L.C.D. $= 12x$

Refer to Study Problem 5 Chapter 8

$$2x=9$$
$$x=4\tfrac{1}{2} \text{ days for Walter to finish the house}$$

60.

	Days to fill reservoir	*Part filled in 1 day*
Storage pipe	15	1/15
Smaller pipe	20	1/20
Together	x	$1/x$
Part to be completed by smaller pipe		13/15

Equation: $\frac{1}{20} = \frac{13}{15x}$

L.C.D. $= 60x$

Refer to Study Problem 7 Chapter 8

$$3x = 52$$
$x = 17\frac{1}{3}$ days for smaller pipe to finish filling the reservoir

61. Let $x =$ price per ounce in 1972

Refer to Example 1 Chapter 3

$$4\frac{1}{2}x = 900$$
$$4.5x = 900$$
$x = \$200$ per ounce in 1972

62. Let $x =$ present office space
$1\frac{1}{2}x =$ increase in office space

Refer to Example 1 Chapter 3

$$x + 1\frac{1}{2}x = 500{,}000$$
$$2\frac{1}{2}x = 500{,}000$$
$x = 200{,}000$ sq ft presently
$1\frac{1}{2}x = 300{,}000$ sq ft to be added

63. Let $x =$ cars sold in 1979
$y =$ decline in car sales in 1982

Refer to Study Problem 7 Chapter 15

Set up two equations:

(1) $y = x/6 - 3{,}000$
(2) $x - y = 300{,}000$

Change (1) to standard form:

$$y = x/6 - 3{,}000$$
$$6y = x - 18{,}000$$
$$-x + 6y = -18{,}000$$

Add (1) and (2):

$$\begin{array}{r} -x + 6y = -18{,}000 \\ x - y = 300{,}000 \\ \hline 5y = 282{,}000 \end{array}$$
$y = 56{,}400$ decline in car sales in 1982

Substitute 56,400 for y in (2)

$$x - 56{,}400 = 300{,}000$$
$$x = 300{,}000 + 56{,}400$$
$x = 356{,}400$ cars sold in 1979

64. Let $x =$ number of cars presently on the road in L.A.
$\frac{1x}{5} - 10{,}000 =$ number of cars that should be reduced

Refer to Study Problem 7 Chapter 15

$$x - \left(\frac{1x}{5} - 10{,}000\right) = 480{,}000$$

L.C.D. $= 5$

$$5x - (1x - 50{,}000) = 2{,}400{,}000$$
$$5x - 1x + 50{,}000 = 2{,}400{,}000$$
$$4x = 2{,}400{,}000 - 50{,}000$$
$$4x = 2{,}350{,}000$$

$$
\begin{aligned}
x &= 587{,}500 \text{ cars now on the road}\\
&\underline{-480{,}000} \text{ should be on the road}\\
&107{,}500 \text{ cars to be reduced}
\end{aligned}
$$

65. x, $x+24$

Let $x =$ width

$x + 24 =$ length

$$
\begin{aligned}
\text{Perimeter} &= 2x + 2(x+24)\\
\text{Perimeter} &= 2x + 2x + 48\\
212 &= 2x + 2x + 48\\
212 &= 4x + 48\\
212 - 48 &= 4x\\
164 &= 4x\\
x &= 41 \text{ ft width}\\
x + 24 &= 65 \text{ ft length}
\end{aligned}
$$

Refer to Problem 3 Chapter 3

66.

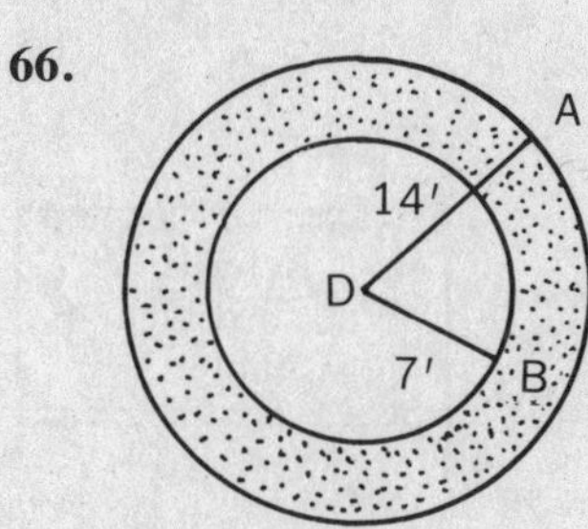

The radius of the larger circle is 14 ft and the radius of the smaller circle is 7 ft. Using the formula $A = \pi r^2$ and $\pi = 22/7$, we have

Refer to Chapter 3

$A = \frac{22}{\not 7} \cdot \not{49}^{\,7} = 154$ sq ft

$A = 154$ sq ft for the smaller circle

$A = \pi r^2$

$A = \frac{22}{\not 7} \cdot \frac{\not{196}^{\,28}}{1} = 616$ sq ft for larger circle

616 sq ft $-$ 154 sq ft $=$ 462 sq ft (area of shaded part of the two concentric circles)

67. $r =$ rate of boat in still water
$x =$ rate of current

Refer to Study Problem 11 Chapter 15

↓ Downstream	↑ Upstream
$d = 50$ miles	$d = 50$ miles
$r = r + x$	$r = r - x$
$t = 2\frac{1}{2}$ hr	$t = 4$ hr

Equation 1	Equation 2
$\frac{50}{r+x} = 2\frac{1}{2}$ hr	$\frac{50}{r-x} = 4$ hr
L.C.D. $= 2(r+x)$	L.C.D. $= (r-x)$
$100 = 5r + 5x$	$50 = 4r - 4x$

(1) $5r + 5x = 100$

(2) $4r - 4x = 50$

Multiply (1) by 4 $\quad 20r + 20x = 400$

Multiply (2) by 5 $\quad \underline{20r - 20x = 250}$

Add (1) and (2) $\quad 40r = 650$

$r = 16\frac{1}{4}$ mph (rate of boat in still water)

Substitute $16\frac{1}{4}$ for r in (1)

$4(16\frac{1}{4}) - 4x = 50$

$x = 3\frac{3}{4}$ mph (rate of current)

68. Equation:

$$60x - 45x = 105$$
$$15x = 105$$
$$x = 7 \text{ quarters}$$
$$2x = 14 \text{ dimes}$$

Refer to Study Problem 1 Chapter 5

69.

	Distance	*Rate*	*Hours*
Federal bus	$56x$	56 mph	x
2nd bus	$60(x - \frac{1}{2})$	60 mph	$(x - \frac{1}{2})$

Equation:

$$56x = 60(x - \tfrac{1}{2})$$
$$56x = 60x - 30$$
$$-4x = -30$$

$x = 7\frac{1}{2}$ hrs travel time for Federal Railways bus

$(x - \frac{1}{2}) = 7$ hrs for second bus to overtake Federal Railways bus

Refer to Study Problem 1 Chapter 4

70.

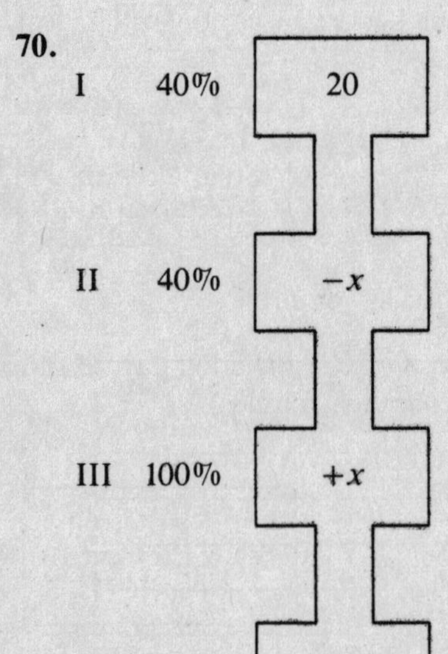

Refer to Study Problem 4 Chapter 6

$$\text{I} - \text{II} + \text{III} = \text{IV}$$
$$40\%(20) - 40\%(x) + 100\% \cdot x = 50\% \cdot (20)$$
$$.40(20) - .40x + 1.00x = .50(20)$$

Move decimal point 2 places to the right.

$$40(20) - 40(x) + 100x = 50(20)$$
$$800 - 40x + 100x = 1{,}000$$
$$60x = 1{,}000 - 800$$
$$60x = 200$$
$$x = 3\tfrac{1}{3} \text{ quarts}$$

$3\frac{1}{3}$ quarts must be drained off and added

71. Let x = age of older man
Let y = age of younger man

Refer to Study Problem 8 Chapter 15

Equations: (1) $x + y = 62$

(2) $\dfrac{y+6}{2} = \dfrac{x+2}{3}$

Convert (2) to standard form

$$3y + 18 = 2x + 4$$
$$-2x + 3y = -14$$

Multiply (1) by 2 $\quad 2x + 2y = 124$

Add (1) and (2) $\quad -2x + 3y = -14$

$$5y = 110$$

$y = 22$ age of younger man

Substitute 22 for y in (1) $\quad x + 22 = 62$

$x = 40$ age of older man

72.

$$6n° = 180°$$
$$n° = 30°$$
$$\text{Angle C} = 30°$$
$$\text{Angle B} = 60°$$
$$\text{Angle A} = 90°$$

Refer to Chapter 3

73.

$$C = \pi d$$
$$154 = 22/7 \cdot d$$
$$49\text{ cm} = d$$

Refer to Chapter 3

74.

	Hours to empty tank		*Part emptied in 1 hr*
Old pump	8		1/8
New pump	4		1/4
Hours for new pump	x		$1/x$
What part completed	3/8	Needs to be done	5/8

Equation: $\frac{1}{4} = \frac{5}{8x}$

L.C.D. $= 8x$

$8x = 20$

$x = 2\frac{1}{2}$ hours for new pump to complete the job

Refer to Study Problem 5 Chapter 8

75.

	Days to build wall		*Work done in 1 day*
Bricklayer	8		1/8
Helper	10		1/10
Days for helper to finish job	x		$1/x$
Part completed	3/8	To be completed	5/8

Equation: $\frac{1}{10} = \frac{5}{8x}$

L.C.D. $= 40x$

$4x = 25$

$x = 6\frac{1}{4}$ days for helper to finish building the wall

Refer to Study Problem 5 Chapter 8

76. Proportion: $\frac{6\text{ ft}}{14} = \frac{60\text{ ft}}{x}$

$6x = 840$

$x = 140$ ft length of the pole's shadow

Refer to Study Problem 2 Chapter 13

77. Proportion: $\frac{7}{4} = \frac{x}{x-45}$

$7x - 315 = 4x$

$3x = 315$

$x = 105$ the larger number

$x - 45 = 60$ the smaller number

Refer to Study Problem 3 Chapter 13

78.

	Age now	*Age in 12 years*
Mr. Santos	$x+28$	$x+40$
Roberto	x	$x+12$

Equation:

$$x+40=2(x+12)$$
$$x+40=2x+24$$
$$-x=-16$$
$$x=16 \text{ age of Roberto now}$$
$$x+28=44 \text{ age of Mr. Santos now}$$

Refer to Study Problem 4 Chapter 11

79.

	Distance	*Rate*	*Time*
With wind	500	$r+x$	2/3
Against wind	500	$r-x$	5/6

Equations: (1) $\dfrac{500}{r+x}=\dfrac{2}{3}$ (2) $\dfrac{500}{r-x}=\dfrac{5}{6}$

$2r+2x=1{,}500$ $5r-5x=3{,}000$

Refer to Study Problem 3 Chapter 13

Multiply (1) by 5 $10r+10x=7{,}500$

Multiply (2) by 2 $10r-10x=6{,}000$

Add (1) and (2) $20r=13{,}500$

$r=675$ km per hr speed of plane

75 km per hr speed of wind

80. Let $x=$ men required

$$80x=100\cdot 40$$
$$80x=4{,}000$$
$$x=50$$

Refer to Study Problem 1 Chapter 3

17 Problems Involving Quadratic Equations

Quadratic equations are equations of the second degree, that is, equations that have an exponent of 2.

Examples

a. $x^2-10x+25=0$ **c.** $15x^2-15x-30=0$

b. $3x^2-8x+5=0$ **d.** $3a^2-4a-7=0$

This chapter will focus on the solution of quadratic equations of *3 terms* by use of the *factoring method*.

Factoring is the reverse process of multiplication. By using the factoring method, each of the above equations will have 2 values for the variable.

a. $(+5, +5)$ **c.** $(+2, -1)$

b. $(1\frac{2}{3}, 1)$ **d.** $(-1, 2\frac{1}{3})$

RULE

If the product of 2 factors is zero, then one of the 2 factors is equal to zero.

Examples

a. If $a \cdot 5=0$
Then $a=0$

b. If $(a+2)\cdot(a-5)=0$
Then $a+2=0$
$a=-2 \longrightarrow -2$
Then $(a-5)=0$
$a=+5 \longrightarrow +5$
} 2 roots

c. $a(a-6)=0$
$a=0$
$a-6=0$
$a=+6$ The 2 roots are $(0, 6)$.

d. $(2m-7)\cdot(3m+4)=0$

$$2m-7=0$$
$$2m=7$$
$$m=3\tfrac{1}{2}$$
$$3m+4=0$$
$$3m=-4$$
$$m=-1\tfrac{1}{3}$$

The 2 roots are $(3\tfrac{1}{2}, -1\tfrac{1}{3})$.

SOLVING QUADRATIC TRINOMIALS BY THE FACTORING METHOD

$(x+5)$ and $(x+4)$ are the factors of the trinomial $x^2+9x+20$.
$(a-7)$ and $(a+5)$ are the factors of $a^2-2a-35$.
$(2m+7)$ and $(m-4)$ are the factors of $2m^2-1m-28$.

Example 1

The procedure for finding the two factors of a quadratic trinomial $x^2-8x+15=0$ is as follows:

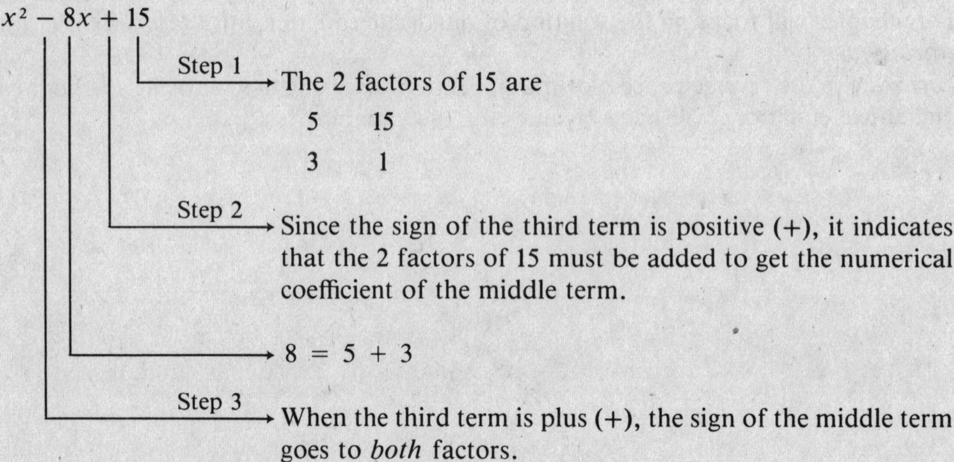

The two factors of $x^2-8x+15$ are $(x-5)$ and $(x-3)$.

Example 2

Using the factoring method to find the 2 roots of the quadratic equation $x^2-7x-18=0$.

$x^2 - 7x - 18 = 0$

Step 1 → The 2 factors of 18 are

6	9	15
3	2	1

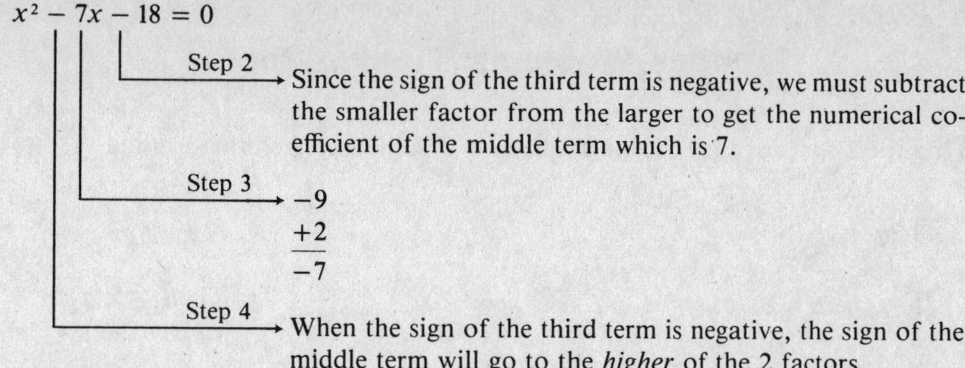

The factors of $x^2 - 7x - 18$ are $(x-9)$ and $(x+2)$.
The two roots are $(9, -2)$.

RULE I

If the sign of the third term is positive, the sign of the middle term goes to *both* factors of the third term.

RULE II

If the sign of the third term is negative, the sign of the middle term goes to the *higher* factor of the third term.

Now we will use the rules for finding the factors of a quadratic equation and the 2 roots.

Example 3

$m^2 - 10m + 24 = 0$

$m^2 - 10m + 24 = 0$

Step 1 → The 2 factors of 24 are

6	12	8	24
4	2	3	1

Step 2 → The sign is (+) in the third term. Therefore, add the 2 factors of 24 which will equal the numerical coefficient of the middle term 10 : 6 and 4.

Step 3, Step 4 → Since the sign of the third term is positive, the sign of the middle term, which is negative, will go to both factors of 24, namely −6 and −4.

The 2 factors of $m^2 - 10m + 24 = 0$

are $(m-6)(m-4) = 0$

$$m - 6 = 0$$
$$m = +6 \leftarrow$$
$$m - 4 = 0$$
$$m = +4 \leftarrow$$

} 2 roots

The 2 roots are $(+6, +4)$.

Practice Problems: Finding Roots I

Find the 2 roots of the following equations by using the factoring method.

1. $a^2-7a+10=0$
2. $r^2+12r+20=6$
3. $r^2-1r-25=0$
4. $t^2+14t-15=0$
5. $y^2-11y-60=0$

Check your answers with the solutions that follow.

Answers to Practice Problems: Finding Roots I

1. $(a-5)(a-2)=0$

$$a-5=0 \qquad a-2=0$$
$$a=5 \qquad a=2$$

The 2 roots are (5, 2).

2. $(r+10)(r+2)=0$

$$r+10=0 \qquad r+2=0$$
$$r=-10 \qquad r=-2$$

The 2 roots are (−10, −2).

3. $(r-8)(r+7)=0$

$$r-8=0 \qquad r+7=0$$
$$r=8 \qquad r=-7$$

The 2 roots are (8, −7).

4. $(t+15)(t-1)=0$

$$t+15=0 \qquad t-1=0$$
$$t=-15 \qquad t=1$$

The two roots are (−15, 1).

5. $(y-15)(y+4)=0$

$$y-15=0 \qquad y+4=0$$
$$y=+15 \qquad y=-4$$

The two roots are (15, −4).

When the numerical coefficient of the x^2 term is greater than 1, we can use the *cross product method* to solve for the 2 roots.

Example 1

$3x^2 - 8x + 5 = 0$

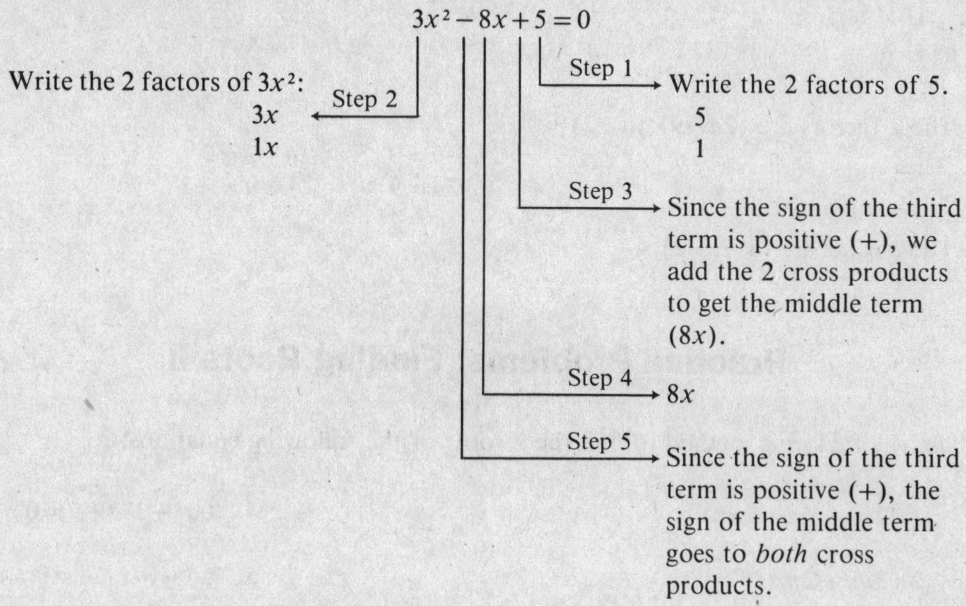

$(3x \quad -5)$ $\qquad 3x \cdot -1 = -3x$ cross product

$(1x \quad -1)$ $\qquad 1x \cdot -5 = -5x$ cross product

The two factors are $(3x-5)(1x-1) = 0$

$$3x - 5 = 0 \qquad 1x - 1 = 0$$

$$x = 1\tfrac{2}{3} \qquad 1x = 1$$

The two roots are $(1\tfrac{2}{3}, 1)$.

Example 2

$2a^2 - 5a - 18 = 0$. Find the 2 roots.

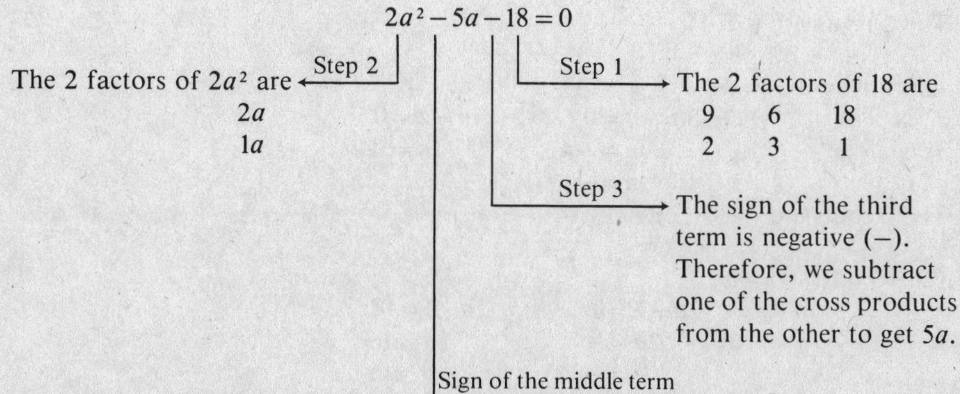

Cross products

$(2a \quad -9)$ $\qquad 1a \cdot -9 = -9a$ cross product

$(1a \quad +2)$ $\qquad 2a \cdot 2 = \underline{\ \ 4a}$ cross product

$$-5a$$

Step 4 → Since the sign of the third term is negative, then the sign of the middle term will go to the higher of the two cross products.

That is,

$$\begin{array}{r} -9a \\ +4a \\ \hline -5a \end{array}$$

The 2 factors are $(2a-9)(1a+2)=0$

$$\begin{aligned} 2a-9&=0 \qquad & 1a+2&=0 \\ 2a&=9 & a&=-2 \\ a&=4\tfrac{1}{2} \end{aligned}$$

The 2 roots are $(4\frac{1}{2}, -2)$.

Practice Problems: Finding Roots II

Use the factoring method to find the 2 roots of the following equations:

1. $2a^2-11a+5=0$
2. $3m^2-2m-8=0$
3. $4a^2+12a-7=0$
4. $10a^2+21a+9=0$
5. $7x^2-10x-8=0$

Check your answers with the solutions that follow.

Answers to Practice Problems: Finding Roots II

1. $(2a-1)(a-5)=0$

$$\begin{aligned} 2a-5&=0 \qquad & a-5&=0 \\ 2a&=1 & a&=+5 \\ a&=\tfrac{1}{2} \end{aligned}$$

The 2 roots are $(\frac{1}{2}, 5)$.

2. $(3m+4)(m-2)=0$

$$\begin{aligned} 3m+4&=0 \qquad & m-2&=0 \\ 3m&=-4 & m&=+2 \\ m&=-1\tfrac{1}{3} \end{aligned}$$

The 2 roots are $(-1\frac{1}{3}, 2)$.

3. $(2a+7)(2a-1)=0$

$$\begin{aligned} 2a+7&=0 \qquad & 2a-1&=0 \\ 2a&=-7 & 2a&=1 \\ a&=-3\tfrac{1}{2} & a&=\tfrac{1}{2} \end{aligned}$$

The 2 roots are $(-3\frac{1}{2}, \frac{1}{2})$.

4. $(5a+3)(2a+3)=0$

$$\begin{aligned} 5a+3&=0 \qquad & 2a+3&=0 \\ 5a&=-3 & 2a&=-3 \\ a&=-\tfrac{3}{5} & a&=-1\tfrac{1}{2} \end{aligned}$$

The 2 roots are $(-\frac{3}{5}, -1\frac{1}{2})$.

5. $(7x+4)(x-2)=0$

$$7x+4=0 \qquad x-2=0$$
$$7x=-4 \qquad x=+2$$
$$x=-4/7$$

The 2 roots are $(-4/7, 2)$.

WORD PROBLEMS INVOLVING QUADRATIC EQUATIONS

STUDY PROBLEM 1

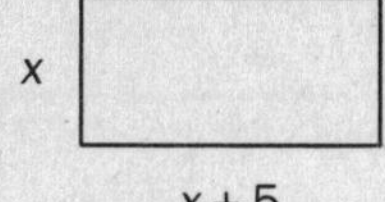

If the length of the rectangle is 5 feet longer than the width and the area is 50 square feet, what are the dimensions of the rectangle? Area = length · width.

Let $x=$ width
$x+5=$ length
$x(x+5)=$ area

Equation:

$$x(x+5)=50 \text{ sq ft}$$
$$x^2+5x-50=0$$
$$(x+10)(x-5)=0$$
$$x+10=0$$
$$x=-10$$
$$x-5=0$$
$$x=+5 \text{ ft}$$

Roots are (5 ft, −10)

However, the length of a line is never negative, so in this instance, the width is 5 feet and the length is 10 feet.

STUDY PROBLEM 2

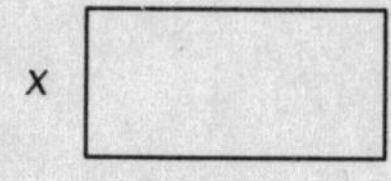

If the area is 55 square feet, find the length and width of the rectangle.

Let $x=$ width
$x+6=$ length

$$x(x+6)=55$$
$$x^2+6x=55$$
$$x^2+6x-55=0$$
$$(x+11)(x-5)=0$$
$$x=5 \text{ ft width}$$
$$x+6=11 \text{ ft length}$$

STUDY PROBLEM 3

One number is 5 times a smaller number. If the product of their squares is 2,600, what are the numbers?

Let $x =$ smaller number
$5x =$ larger number
$x^2 =$ square of smaller number
$25x^2 =$ square of larger number

$$x^2 + 25x^2 = 2{,}600$$
$$26x^2 = 2{,}600$$
$$x^2 = 100$$
$$\sqrt{x^2} = \sqrt{100}$$
$$x = +10 \text{ and } x = -10$$
$$5x = +50 \text{ and } 5x = -50$$

STUDY PROBLEM 4

Solve the following quadratic equation:

$$\frac{1}{x} + \frac{1}{x+2} = \frac{5}{12}$$

L.C.D. $= 12(x)(x+2)$

Clear Fractions:

$$12(\cancel{x})(x+2)\frac{1}{\cancel{x}} + \frac{1}{\cancel{x+2}}12(x)(\cancel{x+2}) = \frac{5}{\cancel{12}}\cancel{12}\cdot x(x+2)$$
$$12(x+2) + 12x = 5x(x+2)$$
$$12x + 24 + 12x = 5x^2 + 10x$$
$$0 = 5x^2 - 14x - 24$$
$$0 = (5x+6)(x-4)$$
$$0 = 5x + 6 \qquad 0 = x - 4$$
$$-6 = 5x \qquad 4 = x$$
$$-1\tfrac{1}{5} = x$$

Roots are $(-1\tfrac{1}{5}, 4)$

STUDY PROBLEM 5

If 48 square inches are subtracted from the area of a square, the result would be equal to twice the perimeter. What is the length of the side of the square?

x square x

$$\text{Area of square} = x^2$$
$$\text{Perimeter} = 4x$$
$$x^2 - 48 = 2(4x)$$
$$x^2 - 48 = 8x$$
$$x^2 - 8x - 48 = 0$$
$$(x-12)(x+4) = 0$$
$$x = +12 \text{ (length of side of square)}$$
$$x = -4$$

STUDY PROBLEM 6

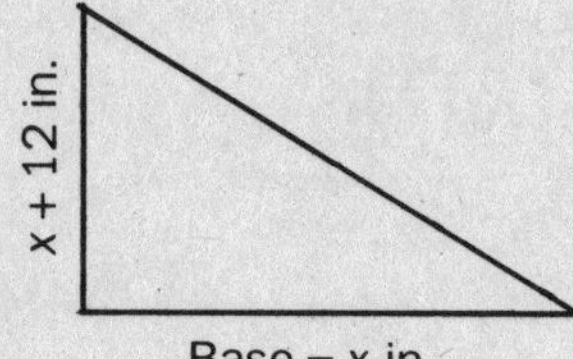

The altitude of the triangle is 12 in. longer than the base. The area of the triangle is A = ½ base · altitude.

Let x = base of triangle
$x+12$ = altitude of triangle

Find the base and the altitude if the area is equal to 32 square inches.

$$A = \tfrac{1}{2}(x)\cdot(x+12)$$
$$32 = \tfrac{1}{2}(x)\cdot(x+12)$$
$$64 = x(x+12)$$
$$64 = x^2 + 12x$$
$$0 = x^2 + 12x - 64$$
$$0 = (x+16)(x-4)$$
$$x = +4 \text{ in. base of the triangle}$$
$$x + 12 = 16 \text{ in. altitude of the triangle}$$

STUDY PROBLEM 7

The area of a rectangle is equal to the area of a square. The length of the rectangle is 6 inches less than twice the side of the square. The width of the rectangle is 4 inches less than the side of the square. What is the length of one side of the square?

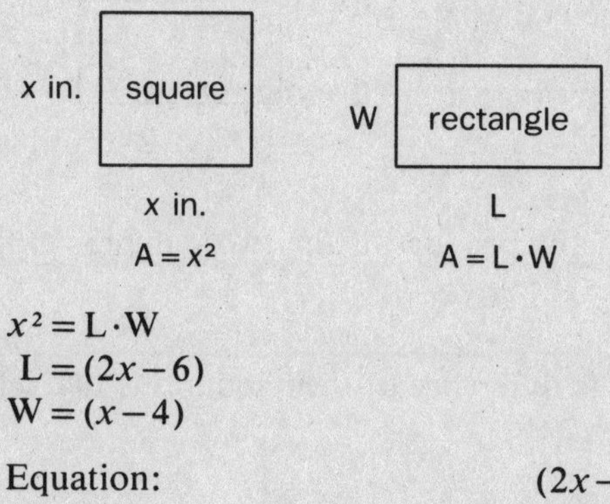

$$x^2 = L \cdot W$$
$$L = (2x-6)$$
$$W = (x-4)$$

Equation:

$$(2x-6)(x-4) = x^2$$
$$2x^2 - 14x + 24 = x^2$$
$$2x^2 - x^2 - 14x + 24 = 0$$

$$x^2 - 14x + 24 = 0$$
$$(x - 12)(x - 2) = 0$$
$$x = +12 \text{ in. length of side of square}$$
$$x = +2$$

STUDY PROBLEM 8

The perimeter of a rectangle is 52 centimeters and the area is 144 square centimeters. What are the dimensions of the rectangle?

Let $x =$ width
$y =$ length
$2x + 2y =$ perimeter
$2x + 2y = 52$

Simplifying by dividing both members by 2, we get

$$x + y = 26$$
$$x = 26 - y$$
$$A = L \cdot W \text{ area of rectangle}$$
$$144 = (26 - y) \cdot y$$
$$144 = 26y - y^2$$
$$y^2 - 26y + 144 = 0$$
factoring $(y - 18)(y - 8) = 0$
$$y - 18 = 0$$
$$y = +18 \text{ length}$$
$$y - 8 = 0$$
$$x = +8 \text{ width}$$

Practice Problems: Quadratic Equations

1. Ronald's age is 8 years greater than Delores's. If the product of their ages is 105 years, what are their ages?

2. If the area of the rectangle is 45 square centimeters, what is the length and the width?

3. One number is larger than another by 15. If the product of the two numbers is 154, what are the 2 numbers?

4. If one number is 6 more than another and the difference of their squares is 144, what are the numbers?

5. The length of a rectangle is 8 feet longer than its width, which has the same dimension as the side of the square. If the area of the rectangle is equal to 160 square feet more than the area of the square, what are the length and width of the rectangle? (Draw diagram.)

6. One holiday, Mr. Anderson drove his family to a picnic ground 60 miles from his house. Returning home that evening, he increased his average speed 9 miles per hour above the rate on the way to the picnic. This increase in miles per hour reduced his return time by 20 minutes. What was his average speed going and returning? (Use model for distance, rate and time.)

7. Working together, an old pump and a new pump can fill a tank in 2 hours. If the new pump can fill the tank in 3 hours less time than the old pump, how many hours will it take each pump to fill the tank? (Use work problem model.)

8. If the area of the triangle is 200 square feet, how many feet are there in BC and AC?

Area of triangle = ½ base · height

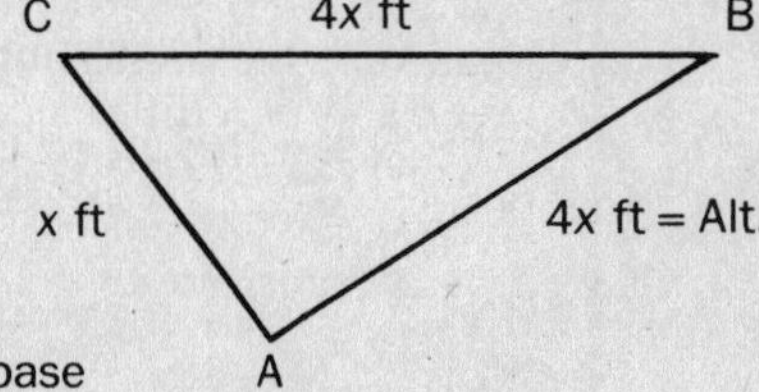

9. The sum of two numbers is 25 and the difference of their squares is 75. What are the 2 numbers?

10. The perimeter of rectangle ABCD is equal to 34 rods. If the area of the rectangle is 60 square rods, what are the dimensions of the rectangle?

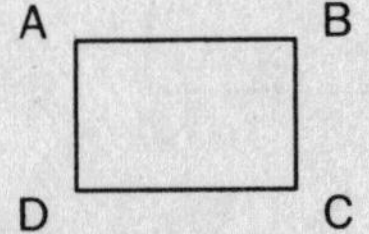

Check your answers with the solutions that follow.

Answers to Practice Problems: Quadratic Equations

1. Let x = Delores's age now
$(x+8)$ = Ronald's age now

$$x(x+8)=105$$
$$x^2+8x-105=0$$
$$(x+15)(x-7)=0 \quad \text{(factoring method)}$$
$$x=-15$$
$$x=7 \text{ Delores's age now}$$
$$x+8=15 \text{ Ronald's age now}$$

2. Let x = width
$(x+12)$ = length
Area $= x(x+12)$

$$45=x^2+12x$$
$$-x^2-12x=-45$$
$$x^2+12x-45=0$$
$$(x+15)(x-3)=0 \quad \text{(factoring method)}$$
$$x=+3$$
$$x=-15$$
$$x=3 \text{ cm}$$
$$x+12=15 \text{ cm}$$

3. Let $x =$ smaller number
$(x+15) =$ larger number

$$x(x+15)=154$$
$$x^2+15x-154=0$$
$$(x+22)(x-7)=0$$
$$x=+7$$
$$x=-22$$

(The 2 numbers are 7 and 22 or −7 and −22.)

4. Let $x =$ smaller number
$(x+6) =$ larger number
$(x+6)^2-x^2 =$ difference of two squares

$$x^2+12x+36-x^2=144$$
$$12x+36=144$$
$$12x=108$$
$$x=9 \text{ smaller number}$$
$$(x+6)=15 \text{ larger number}$$

5.

	square		rectangle
x	(side x)	x	(length x + 8)

$x(x+8) =$ area of rectangle
$x^2 =$ area of square

$$x(x+8)=x^2+160$$
$$x^2+8x=x^2+160$$
$$x^2-x^2+8x=160$$
$$8x=160$$
$$x=20 \text{ ft width of rectangle}$$
$$(x+8)=28 \text{ ft length of rectangle}$$

6.

	Distance	*Rate*	*Time*
Going	60	r	$60/r$
Returning	60	$r+9$	$60/r+9$

$$\frac{60}{r}-\frac{60}{r+9}=\frac{1}{3}$$

L.C.D. $= 3r(r+9)$

Clear fractions:

$$3\not{r}(r+9)\frac{60}{\not{r}}-3r(\not{r+9})\frac{60}{\not{r+9}}=\not{3}r(r+9)\frac{1}{\not{3}}$$
$$180(r+9)-180r=r(r+9)$$
$$180r+1{,}620-180r=r^2+9r$$
$$-r^2-9r+1{,}620=0$$

Changing signs (−): $$r^2+9r-1{,}620=0$$
$$(r+45)(r-36)=0$$
$$r=36 \text{ mph going}$$
$$r=-45$$
$$r+9=45 \text{ mph returning}$$

7.

	Hours to fill tank	*Part filled in 1 hr*
Old pump	x	$1/x$
New pump	$x-3$	$1/x-3$
Together	2	1/2

Equation: $\frac{1}{x}+\frac{1}{x-3}=\frac{1}{2}$

L.C.D. $=2\cdot x\cdot(x-3)$

Clear fractions: $2\cdot \cancel{x}(x-3)\frac{1}{\cancel{x}}+2\cdot x\cancel{(x-3)}\frac{1}{\cancel{x-3}}=2\cdot x(x-3)\frac{1}{2}$

$$2(x-3)+2\cdot x=x^2-3x$$

$$2x-6+2x=x^2-3x$$

Changing signs (–): $-x^2+7x-6=0$

$$x^2-7x+6=0$$

$$(x-6)(x-1)=0$$

$$x=6$$

$$x=1$$

6 hours for old pump to fill the tank
3 hours for new pump to fill tank

8.

$$A=\frac{1}{2}\text{base}\times\text{height}$$

$$200=\frac{1}{\cancel{2}}x\cdot{}^{2}\cancel{4}x^2$$

$$200=2x^2$$

$$100=x^2$$

$10=x$ feet in base AC

$40=4x$ feet in altitude AB

9. Let $x=$ one number
$25-x=$ other number

Equation: $(25-x)^2-x^2=75$

$$625-50x+x^2-x^2=75$$

$$-50x=75-625$$

$$-50x=-550$$

$x=11$ smaller number

$25-x=14$ larger number

10.

$$\text{Perimeter}=2L+2W$$

$$34=2L+2W$$

Simplifying, divide by (2): $17=L+W$

$$17-W=L$$

$$\text{Area of rectangle}=L\cdot W$$

Simplifying by changing signs (–):

$$W(17-W)=60$$

$$-W^2+17W-60=0$$

$$W^2-17W+60=0$$

$$(W-12)(W-5)=0$$

$W=12$ rods: $17-W=L$, then Length $=5$ rods

$W=5$ rods: $17-W=L$, then Length $=12$ rods

18 Trigonometric Word Problems

Trigonometry is the branch of mathematics that deals with the relations between the sides and angles of triangles.

Problems involving *right* triangles can be solved by using the *Pythagorean Theorem*. It has been proven in geometry that in any right triangle, the sum of the squares of each leg is equal to the square of the hypotenuse. This law is called the *Pythagorean Theorem.*

Example 1

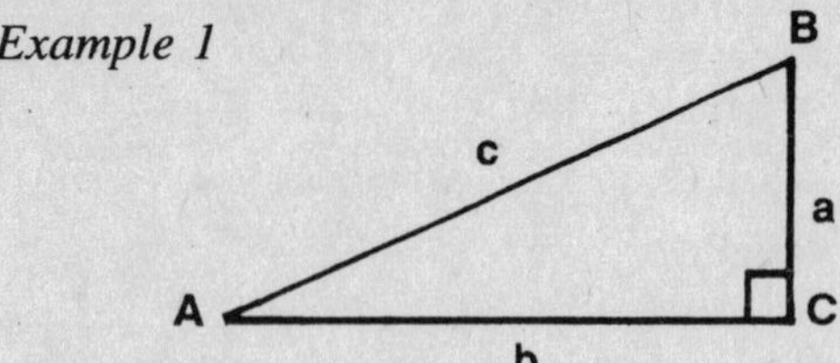

In triangle ABC, angle C is a right angle (90°), the side opposite angle C is called the hypotenuse, and the two other sides are called the legs of the right triangle.

The Pythagorean Theorem states that THE SQUARE OF THE LENGTH OF THE HYPOTENUSE IS EQUAL TO THE SUM OF THE SQUARES OF THE LENGTH OF THE LEGS. (a) and (b) are the two legs and (c) is the hypotenuse. Therefore, the Pythagorean equation is:

$$a^2+b^2 = c^2$$

If (a) = 3 and (b) = 4 and (c) = 5, then:

$$3^2+4^2=5^2$$
$$9+16=25$$
$$25=25$$

Example 2

In triangle ABC, angle A + angle B = 90°. If side a = 12 and side b = 9, what is the length of the hypotenuse?

If angle A + angle B = 90°, then angle C is also 90° because the **sum** of the angles of a triangle is 180° and triangle ABC is a right triangle. Use the Pythagorean Thorem to find the hypotenuse: $a^2+b^2=c^2$

(Refer to Table I at the back of this book to find square roots.)

$$12^2 + 9^2 = C^2$$
$$144 + 81 = C^2$$
$$225 = C^2$$
$$\sqrt{225} = \sqrt{C^2}$$
$$15 = C$$

Note: Although the square root of 225 is also -15, we use the positive number because you can not have a negative side to a triangle.

Example 3

One side of a right triangle is 7 feet shorter than the other. If the hypotenuse is 17 feet long, what is the perimeter of the triangle?

Let x represent the second side.
Let $x-7$ represent the first side.

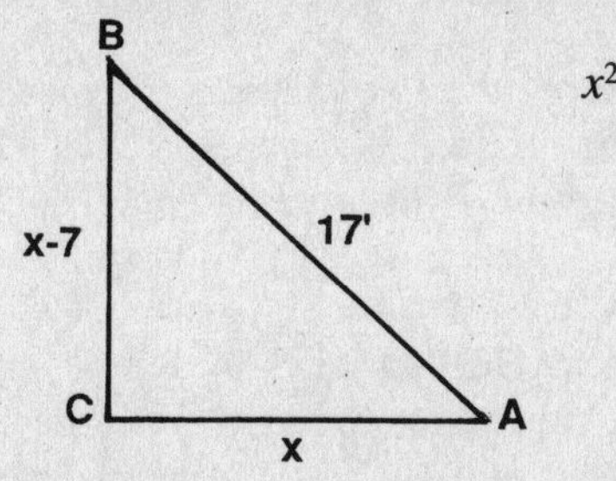

$$x^2 + (x-7)^2 = 17^2$$
$$x^2 + x^2 - 14x + 49 = 289$$
$$2x^2 - 14x - 240 = 0$$
$$(2x-30)(x+8) = 0$$
$$2x - 30 = 0$$
$$2x = 30$$
$$x = 15$$
$$x + 8 = 0$$
$$x = -8$$

We reject the negative number.
The perimeter = 15 + 8 + 17 = 40 feet.

Practice Problems: Right Triangles

1. The length of a rectangle is 5 feet longer than the width. If the diagonal is 25 feet, what are the dimensions of the sides?

2. The sides of a right triangle are consecutive even numbers. Find the length of the sides.

3. The sides of a right triangle differ by 3. Find the sides.

4. One side of a right triangle is 2 inches longer than the other and 2 inches less than the hypotenuse. What are the dimensions of the hypotenuse and the sides?

5. A brace wire extends from the top of a 30 foot pole to a point on the ground 38 feet from the foot of the pole. Allowing 4 feet for attaching the wire, how long is the wire?

Answers to Practice Problems: Right Triangles

1. Let $x =$ the width
$x+5 =$ the length

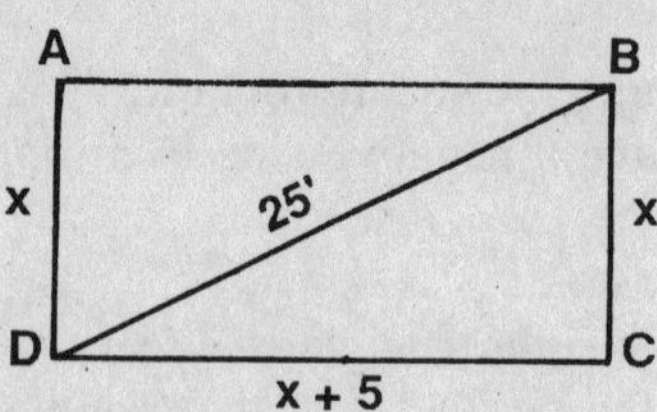

$$(x)^2+(x+5)^2=25^2$$
$$x^2+x^2+10x+25=625$$
$$2x^2+10x-600=0$$
$$x^2+5x-300=0$$
$$(x+20)(x-15)=0$$
$$x=-20 \text{ (reject the negative number)}$$
$$x=+15$$

The dimensions are: width $= x = 15'$
length $= x+5 = 20'$

2. Let $x =$ length of the smallest leg
$x+2 =$ length of other leg
$x+4 =$ length of the hypotenuse

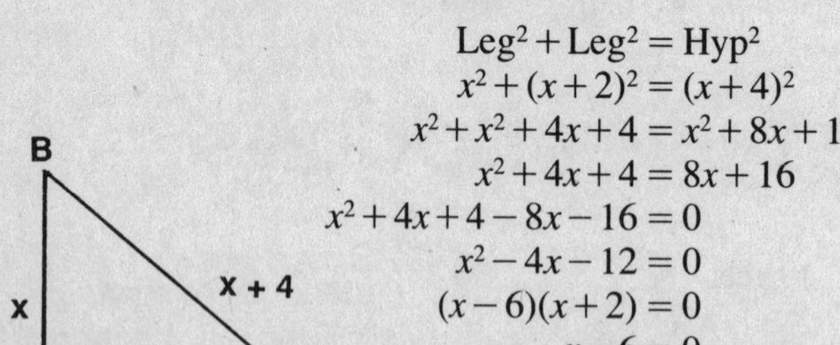

$$\text{Leg}^2+\text{Leg}^2=\text{Hyp}^2$$
$$x^2+(x+2)^2=(x+4)^2$$
$$x^2+x^2+4x+4=x^2+8x+16$$
$$x^2+4x+4=8x+16$$
$$x^2+4x+4-8x-16=0$$
$$x^2-4x-12=0$$
$$(x-6)(x+2)=0$$
$$x-6=0$$
$$x=6$$
$$x+2=0$$
$$x=-2 \text{ (reject the negative number)}$$

Therefore $x = 6$ (smallest leg)
$x+2 = 8$ (other leg)
$x+4 = 10$ (hypotenuse)

3. Let $x =$ length of the smallest leg
$x+3 =$ length of the other leg
$x+6 =$ length of the hypotenuse

$$\text{Leg}^2+\text{Leg}^2=\text{Hyp}^2$$
$$x^2+(x+3)^2=(x+6)^2$$
$$x^2+x^2+6x+9=x^2+12x+36$$
$$x^2+6x+9-12x=36$$
$$x^2-6x-27=0$$
$$(x-9)(x+3)=0$$
$$(x-9)=0$$
$$x=9$$
$$(x+3)=0$$
$$x=-3 \text{ (reject the negative number)}$$

Therefore $x = 9$ (smallest leg)
$x+3 = 12$ (other leg)
$x+6 = 15$ (hypotenuse)

4. Let x = one side
$x-2$ = other side
$x+2$ = hypotenuse

$$x^2+(x-2)^2=(x+2)^2$$
$$x^2+x^2-4x+4=x^2+4x+4$$
$$x^2-4x=4x$$
$$x^2-8x=0$$
$$x(x-8)=0$$
$$x=0$$
$$x-8=0$$
$$x=8$$
$$x-2=6$$
$$x+2=10$$

5. Let x represent the length AB of the wire.

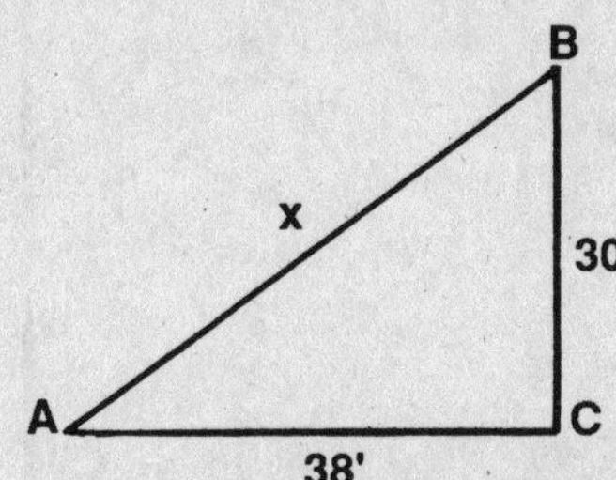

$$x^2=(38)^2+(30)^2$$
$$x^2=1{,}444+900$$
$$x^2=2{,}344$$
$$\sqrt{x^2}=\sqrt{2{,}344}$$
$$x=48.4 \text{ ft}$$
$$48.4+4=52.4 \text{ length of wire}$$

TRIGONOMETRIC FUNCTIONS

To solve some problems involving right triangles, as well as other kinds of triangles, you need to be acquainted with the trigonometric functions that define the ratios between the angles and sides of the triangle. These three functions are called:

SINE
COSINE
TANGENT

An imaginative student came up with a scheme to remember the definitions of the three trigonometric functions. He devised the word: SOHCAHTOA

SOH: Sine = Opposite Leg : the Hypotenuse
CAH: Cosine = Adjacent Leg : the Hypotenuse
TOA: Tangent = Opposite Leg : the Adjacent Side

STUDY PROBLEM 1

How long must a ramp be to reach across a stream 17.2 feet wide, if the angle of elevation with the opposite bank is 22°?

In the figure, we know the angle and the adjacent leg, and wish to find the hypotenuse. We need to know, therefore, the ratio that compares both A (adjacent leg) and

H (the Hypotenuse). Looking at the information given, we know that the COSINE—that is, the relationship between the opposite leg and the hypotenuse—is 22°. Using the COSINE then, we can solve for the hypotenuse.

Angles can be translated into decimals. Fortunately for the student of trigonometry, the relationships between sines, cosines, and tangents have already been compiled and can be found in TABLE 2 at the end of this book.

Examples from TABLE 2

Angle	**Sine**	**Cosine**	**Tangent**
22°	0.375	0.927	0.404
23°	0.397	0.921	0.424

Using SOH*CAH*TOA, we obtain the equation:

$$\cos = \frac{adj.\ leg}{Hyp}$$

$$\cos = \frac{17.2}{AB} \quad \text{(hypotenuse)}$$

Using TABLE 2, substitute the corresponding decimal for 22°.

$$\frac{0.927}{1} = (1)(17.2)$$

$$AB = \frac{17.2}{0.927}$$

$$AB = 18.55$$

$$AB = 18.6 \text{ feet}$$

STUDY PROBLEM 2

How high up a wall does a 10 foot ladder reach, if it has an angle of elevation of 63°?

Draw the picture

10'
x
opp
63°
Adj

Find the ratio that includes the hypotenuse and the opposite leg.

That ratio is SOH or $\text{Sin} = \frac{\text{opposite leg}}{\text{hypotenuse}}$

$$\text{Sin } 63° = \frac{x}{10}$$

Substitute the value for sin 63° from TABLE 2 and solve for x (opposite leg):

$$.891 = \frac{x}{10}$$

$$8.91 = x$$

STUDY PROBLEM 3

From a plane flying at an altitude of 2,000 feet, the angle of depression (the angle looking down) to an airport tower is 62°. Find, to the nearest foot, the distance the plane must fly to be directly over the tower assuming there is no change in altitude.

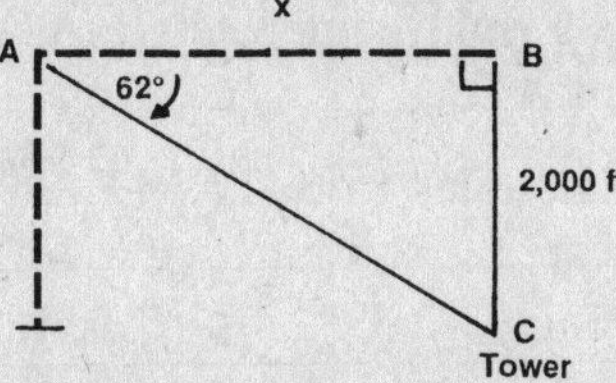

The angle of depression is the angle formed by the plane's path (AB) and the sightline to the tower (AC). Draw a right triangle whose legs are the plane's path and the altitude of the plane directly above the tower, and whose hypotenuse is the sightline to the tower.

You have an angle and the opposite leg and you want to find the adjacent leg. Therefore use:

$$\text{Tangent} = \frac{\text{opposite leg}}{\text{adjacent leg}}$$

$$\text{Tan } 62° = \frac{2000}{x}$$

Substitute value of tan 62° from Table 2 and solve for x.

$$1.881 = \frac{2000}{x}$$

$$x = \frac{2000}{1.881}$$

$$x = 1063 \text{ ft.}$$

Practice Problems: Trigonometric Functions

1. A cable for a 15-foot pole is attached at a 60° angle to the ground. Find the length of the cable to the nearest tenth of a foot.

2. A ladder leans against a chimney. The angle the ladder makes with the ground is 28°. The length of the ladder is 20 feet. Find the distance in feet from the foot of the ladder to the foot of the chimney.

3. A road 700 feet long runs up to the top of the hill with an incline of 20°. What is the height of the hill?

4. An engineer surveying a building set up his transit 100 feet from the base of the building. The angle of elevation with the top of a building was 61°. What is the height of the building?

5. At a point on the ground 50 feet from the foot of a tree, the angle of elevation to the top of the tree is 38°. How high is the tree?

6. From an airplane flying at an elevation of 4,000 feet over the ocean, the angle of depression with the coast of an island is 15°. How many miles must the airplane still go to be directly over the island?

7. From the top of a lighthouse 200 feet high, the angle of depression of a boat at sea measures 36°. Find, to the nearest foot, the distance from the boat to the foot of the lighthouse.

8. In triangle ABC, angle C is a right angle. Angle A is 23° and the hypotenuse is 44 inches. What is the length of AC?

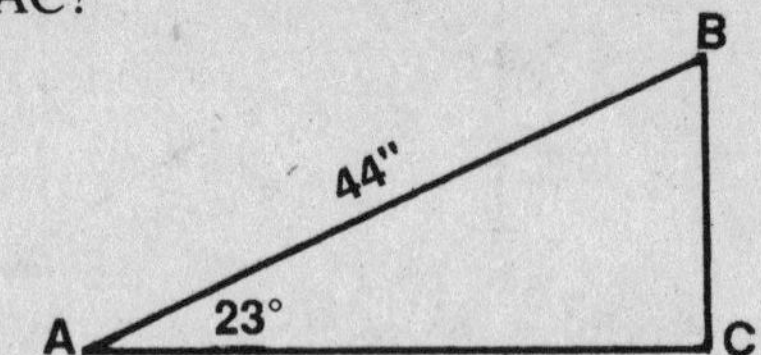

Answers To Practice Problems: Trigonometric Word Problems

1.

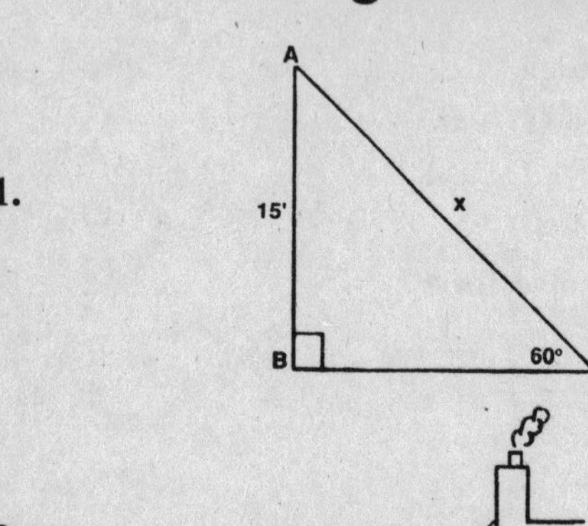

$$\text{Sin } 60° = \frac{15}{x}$$
$$.866 \cdot x = 15$$
$$x = \frac{15}{.866}$$
$$x = 17.3 \text{ ft}$$

2.

$$\text{Cos } 28° = \frac{b}{20}$$
$$20\ (\text{Cos } 28°) = b$$
$$20\ (.883) = b$$
$$b = 17.7 \text{ ft}$$

3.

$$\text{Sin } 20° = \frac{x}{700}$$
$$.342 = \frac{x}{700}$$
$$700\ (.342) = 239.4$$
$$x = 239.4 \text{ ft}$$

4.

$$\text{Tan } 61° = \frac{x}{100}$$
$$1.804 = \frac{x}{100}$$
$$180.4 \text{ ft.} = x$$

5.

$$\text{Tan } 38° = \frac{\text{opp. leg}}{\text{adj. leg}}$$
$$\text{Tan } 38° = \frac{x}{50}$$
$$50\ (.781) = x$$
$$39.05 \text{ ft} = x$$

6.

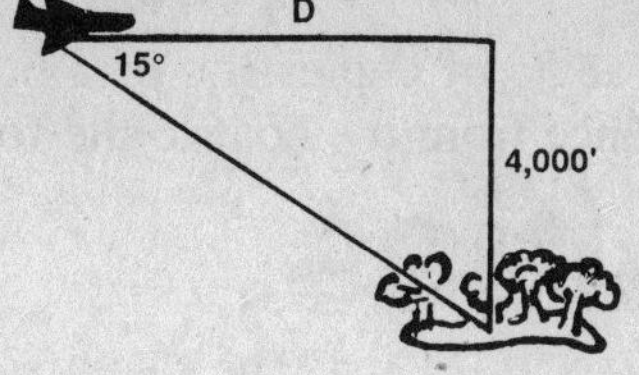

$$\text{Tan } 15° = \frac{4000}{D}$$

$$.268 = \frac{4000}{D}$$

$$D = \frac{4000}{.268}$$

$$D = 14{,}925 \text{ ft}$$

7.

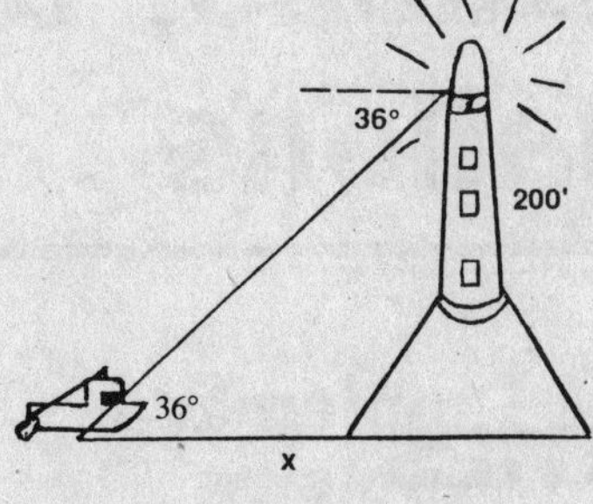

$$\text{Tan } 36° = \frac{200}{x}$$

$$x\,(.727) = 200$$

$$x = \frac{200}{.727}$$

$$x = 275 \text{ ft}$$

8.

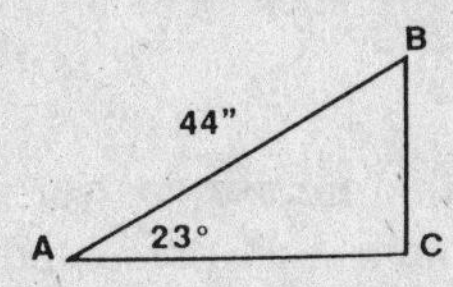

$$\text{Cos } 23° = \frac{\text{AC}}{44}$$

$$(.921)(44) = \text{C}$$

$$40.5'' = \text{AC}$$

19 Probabilities, Permutations, and Combinations

PROBABILITIES

In the previous chapters of this book, each problem had a precise answer. That answer remained the same no matter how many times the problem was presented. There are, however, many problems in which the answers are uncertain. Nonetheless, some answers are *more probable* than others.

Many industries find it useful to determine which answers or solutions are more probable than others. Some auto insurance companies, for example, base some rates on the probability that men and women under 25 years of age will have the most accidents. A gambler in Las Vegas places a bet based on the probability that the number 7 will turn up on the dice before the number 3. Engineers launch space telescopes based on a high probability that the mechanism will function perfectly. A doctor prescribes a particular medication based partially on the probability that there will be few negative side effects. Probability is fundamental to many fields, including business, physics, politics, and the military.

It is possible to set up equations to reflect probabilities.

STUDY PROBLEM 1

From a bag containing twelve white balls and 18 red ones, one ball is drawn at random.

(a) What is the probability that the ball drawn is white?
(b) What is the probability that the ball drawn is red?

12 = the white balls
18 = the red balls
30 = total number of balls

(a) To find the probability of drawing a white ball divide the white balls by the total number of balls and reduce the fraction to the lowest common denominator.

$$12/30 = 2/5$$

Answer: There is a $2/5$ probability that the ball drawn is white.

(b) To find the probability of drawing a red ball divide the red balls by the total number of balls and reduce the fraction to the lowest common denominator.

$$^{18}/_{30} = {}^{3}/_{5}$$

Answer: There is a $^{3}/_{5}$ probability that the ball drawn is red. It can also be said that there is a $^{1}/_{5}$ greater probability that the ball drawn will be red.

SAMPLE SPACES

Toss a coin 30 times. Does your experiment show that heads comes up $^{15}/_{30}$ or $^{1}/_{2}$ the time? If not, remember that we are only talking about probabilities, not exact outcomes. A *sample space* for this experiment is (H, T).

There are two possible *events* when you toss a coin: heads (H) and tails (T). When an event involves a single element of the sample space, it is called a *simple event*. (H) and (T) are both simple events.

Either one of the two outcomes H or T is equally likely to occur. The probability of the simple event (H), then, is the same as the probability of (T). Therefore:

$$P(H) = P(T) = {}^{1}/_{2}$$

Since there are two possible outcomes, one for Heads and one for Tails, then P = 2. The probability, then, that heads will turn up is:

$$HP \text{ or } {}^{1}/_{2}$$

The probability that T will turn up is:

$$TP \text{ or } {}^{1}/_{2}$$

Therefore, the *chances are* one in two or one-half that a head will turn up in one toss. If you toss the coin 20 times, the probability is $^{10}/_{20}$ or still one-half.

STUDY PROBLEM 2

Two dice (with numbers 1 through 6) are rolled and the numbers are recorded. Find the probability of each event below:

Event *A:* (the sum is 2 or 3)
Event *B:* (the sum is greater than 8)

Solution: First we must define the *sample spaces* (all the possible outcomes) for rolling two dice.

$$\begin{aligned} S = {} & (1, 1)\ (1, 2)\ (1, 3)\ (1, 4)\ (1, 5)\ (1, 6)\ (2, 1)\ (2, 2)\ (2, 3)\ (2, 4)\ (2, 5)\ (2, 6) \\ & (3, 1)\ (3, 2)\ (3, 3)\ (3, 4)\ (3, 5)\ (3, 6)\ (4, 1)\ (4, 2)\ (4, 3)\ (4, 4)\ (4, 5)\ (4, 6) \\ & (5, 1)\ (5, 2)\ (5, 3)\ (5, 4)\ (5, 5)\ (5, 6)\ (6, 1)\ (6, 2)\ (6, 3)\ (6, 4)\ (6, 5)\ (6, 6) \end{aligned}$$

The number (n) of ordered pairs in the above set is:

$$n(S) = 36$$

The number (n) of ordered pairs that add up to 2 or 3 are:

Event $A = (1, 1)\ (1, 2)\ (2, 1)$
or $n(A) = 3$

The number of ordered pairs that total more than 8 are:

Event B = (6, 3) (6, 4) (6, 5) (6, 6) (5, 4) (5, 5) (5, 6) (4, 5) (4, 6) (3, 6)
or $n(B)$ = 10

The probability for Event A is the number of ordered sets in Event A divided by the number of ordered sets in the *sample space*.

$$P\ (A) = \frac{n(A)}{n(S)} = \frac{3}{36} = \frac{1}{12}$$

The probability for Event B is the number of ordered sets in Event B divided by the number of ordered sets in the *sample space*.

$$P\ (A) = \frac{n(B)}{n(S)} = \frac{10}{36} = \frac{5}{18}$$

Practice Problems: Probabilities

1. Two coins are tossed. Find the probability that both show heads.
2. Two dice are rolled. What is the probability that the sum will be 7?
3. One marble is drawn from a bag containing two red, two black, and two green marbles. What is the probability that a red marble will be drawn?

Answers to Practice Problems: Probabilities

1. $S = (H, H)\ (H, T)\ (T, H)\ (T, T)$
   Event $A = (H, H)$

$$P\ (A) = \frac{n(A)}{n(S)} = \frac{1}{4}$$

The probability is ¼ that both coins will turn up heads.

2. S = (1, 1) (1, 2) (1, 3) (1, 4) (1, 5) (1, 6) (2, 1) (2, 2) (2, 3) (2, 4) (2, 5) (2, 6)
(3, 1) (3, 2) (3, 3) (3, 4) (3, 5) (3, 6) (4, 1) (4, 2) (4, 3) (4, 4) (4, 5) (4, 6)
(5, 1) (5, 2) (5, 3) (5, 4) (5, 5) (5, 6) (6, 1) (6, 2) (6, 3) (6, 4) (6, 5) (6, 6)

The number (n) of ordered pairs in the above set is:

$$n(S) = 36$$

The number (n) of ordered pairs that add up to 7 are:

Event A = (1, 6) (2, 5) (3, 4) (4, 3) (5, 2) (6, 1)
or $n(A) = 6$

Divide the number of ordered sets in Event A by the number of ordered sets in the Sample Space:

$$P(A) = \frac{n(A)}{n(S)} = \frac{6}{36} = \frac{1}{6}$$

3. S = (r) (r) (b) (b) (g) (g)
Event A = (r) (r)
Event $A = 2$

$$P(A) = \frac{n(A)}{n(S)} = \frac{2}{6} = \frac{1}{3}$$

Permutations and Combinations

Although probabilities are imprecise, the *number of possible outcomes* is precise.

Example 1

If you have four sweaters and two pairs of jeans, how many different sweater and jean combinations can you make?

Let A, B, C, and D represent the sweaters and 1 and 2 the pairs of jeans. We can represent these 8 combinations on the diagram below.

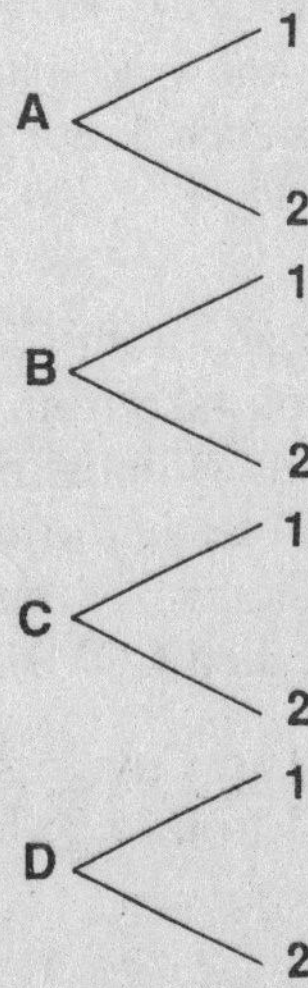

Since there are four sweaters and two pairs of jeans from which to choose, there are 4 x 2 = 8 possible combinations.

In other words, if you can select one item in "X" different ways and then another selection in "Y" different ways, the number of ways the two selections can be made is X × Y different ways.

Example 2

What are the possible outcomes if three coins, a nickel, dime and quarter, are tossed?

H = heads
T = tails

	5	**10**	**25**
1.	H	H	H
2.	H	H	T
3.	H	T	H
4.	H	T	T
5.	T	H	H
6.	T	H	T
7.	T	T	H
8.	T	T	T

As you can see by the chart, there are eight different possible outcomes. Since each coin can turn up two possible ways, then:

$$2 \times 2 \times 2 = 8$$

Example 3

In an election, there are three candidates for senator and four for governor. How many ways can a ballot be marked for both of these offices?

This problem involves finding the number of different ways two acts can be performed in succession. If one action can be done in any one of n different ways, and if, after it has been done, a second action can be done in any one of m different ways, then both acts can be performed in the order stated in nm different ways.

The ballot can be marked for senator in any one of (3) ways. With any one of these (3) ways, we can associate any one of (4) ways. Thus, the number of ways a ballot can be marked for both offices is $3 \times 4 = 12$ different ways.

If we wanted to know how many ways three ballots could be marked for both offices, we would multiply each ballot times the number of possible combinations.

Example 4

There are four roads from City 1 to City 2 and five roads from City 2 to City 3. How many routes can you take from City 1 via City 2 to City 3?

The question deals with the number of ways each element of one set (A, B, C) can be matched with each element of another set (E, F, G, H, I).

Ordered arrangements of numbers are called PERMUTATIONS. Subsets of these arrangements are called COMBINATIONS.

Routes from City 1 to City 2: A, B, C, D
Routes from City 2 to City 3: E, F, G, H, I

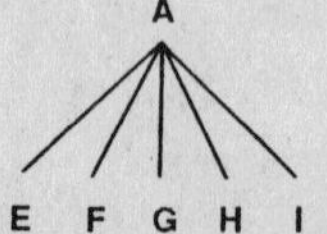

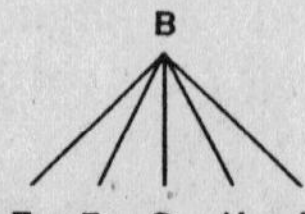

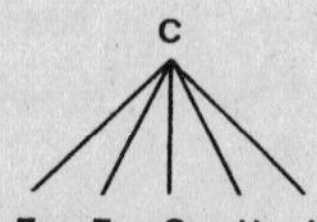

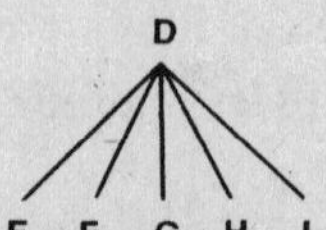

4 roads × 5 roads = 20 routes to City 3

Example 5

How many 3-digit, positive, even integers are there?

The number is even. Therefore, it must end with either a 0, 2, 4, 6 or 8. There are five different choices for the last digit.

___ ___ <u>5</u>

For the number to be meaningful, it must start with either a 1, 2, 3, 4, 5, 6, 7, 8 or 9.

<u>9</u> ___ <u>5</u>

The second digit may be any of the digits 0, 1, 2, 3, 4, 5, 6, 7, 8 or 9. There are ten possible numbers for the second digit.

<u>9</u> <u>10</u> <u>5</u>

There are 9 x 10 x 5 or 450 different 3-digit positive, even integers.

Practice Problems

1. If a girl has six different skirts and ten different blouses, how many different skirt-and-blouse outfits can she put together?
2. At a dance, there are 25 boys and 22 girls. How many boy-girl couple combinations are possible?
3. A drawer contains ten spoons, six knives, and five forks. In how many ways can you choose a place setting of a knife, fork and spoon?
4. In how many ways can you answer ten true–false questions?
5. How many two-or three-digit numbers can be formed from the digits 1, 2, 3, 4 and 5, if there are no repeated digits in the number?

Answers to Practice Problems

1. 6 skirts $\times$ 10 blouses $=$ 60 different combinations.
2. 25 boys $\times$ 22 girls $=$ 550 couple combinations
3. $10 \times 6 \times 5 = 300$
4. 10 questions for two ways:

$$2 \times 2 \times 2 \times 2 \times 2 \times 2 \times 2 \times 2 \times 2 \times 2 \text{ or}$$

$$2^{10} = 1{,}024$$

5. If one of the five numbers is the first digit of a 3-digit number, then only four are left for the second digit and three for the third:

$$5 \times 4 \times 3 = 60$$

If one of the five numbers is used in the two-digit number, then only four are left for the second digit.

$$5 \times 4 = 20$$

$$5 \times 4 \times 3 = 60$$
$$5 \times 4 = 20$$

$20+60=$*80* two and three-digit numbers can be formed from the numbers 1, 2, 3, 4 and 5.

PERMUTATIONS

Each arrangement or ordering of a set of objects is called a *permutation* of the set. In the set [1, 2, 3] there are six possible *permutations*.

They are: (1, 2, 3), (1, 3, 2), (2, 1, 3), (2, 3, 1), (3, 1, 2), (3, 2, 1)

There are three possible choices for the first position, either 1, 2, or 3. But once that position is filled, there are only two choices left for the second position, and once that position is filled, only one choice left for the final position, so that:

$$P \text{ (permutations)} = 3 \times 2 \times 1 = 6$$

In other words, the number of permutations possible in a set of n objects is $n(n-1)$ $(n-2)$ $(n-3)$, etc., or *all the whole numbers from one up to that number.*

To indicate that we want to multiply together all the numbers up to an n number, we write the number with an exclamation sign after it.

Example

$$7! = 7 \times 6 \times 5 \times 4 \times 3 \times 2 \times 1$$

So, when we see 7!, we know it represents the product of all the numbers from one to seven. This representation is called the *factorial function.*

The *factorial* of a whole number n is $n!$ The *factorial* 7! is 5,040.

$$7! = 7 \times 6 \times 5 \times 4 \times 3 \times 2 \times 1 = 5{,}040.$$

STUDY PROBLEM 3

How many ways can eight people line up in a cafeteria line?

If the first person is in the first position, then only seven people are left to line up in the second position, six people in the third position, etc.

$$P = 8!$$

$$8 \times 7 \times 6 \times 5 \times 4 \times 3 \times 2 \times 1 = 40{,}320$$

Exercise

Solve the following:

1. $5! =$

2. $(3!)\ (4!) =$

3. $\frac{7!}{5!} =$

4. How many ways can the letters in the word SUNDAY be arranged?

Answers

1. $5 \times 4 \times 3 \times 2 \times 1 = 120$

2. $(3 \times 2 \times 1)\ (4 \times 3 \times 2 \times 1) = (6)\ (24) = 144$

3. $\dfrac{7 \times 6 \times \cancel{5} \times \cancel{4} \times \cancel{3} \times \cancel{2} \times \cancel{1}}{\cancel{5} \times \cancel{4} \times \cancel{3} \times \cancel{2} \times \cancel{1}} = 42$

4. $6 \times 5 \times 4 \times 3 \times 2 \times 1 = 720$

In Example 1 concerning sweaters and jeans, we are not interested in the order in which the sweater and jean combinations are chosen, but only in the number of possible ways they can be chosen.

However, in some situations involving choices, the order in which the choices are made *IS* important.

STUDY PROBLEM 4

Six books are selected from a group of ten and arranged on a shelf. How many arrangements are possible?

Solution: Find the number of permutations of ten taken six at a time.

The symbol *nPr* is used to indicate the number of permutations of n objects taken r at a time.

$${}_{10}P_{6}$$

The factorial of $10 = 10 \times 9 \times 7 \times 6 \times 5 \times 4 \times 3 \times 2 \times 1$ but we will select only six, so:

$${}_{10}P_{6} = 10 \times 9 \times 8 \times 7 \times 6 \times 5 = 151{,}200 \text{ possible arrangements.}$$

STUDY PROBLEM 5

How many different arrangements (permutations) that are distinguishable can be formed from the word LITTLE using all the letters?

There are 6! number of permutations. But since there are two T's and two L's which are repeated and not distinguishable from each other, we must divide by 2! (L's) and 2! (T's).

$$P = \frac{6!}{2!2!} = \frac{6 \cdot 5 \cdot 4 \cdot 3 \cdot 2 \cdot 1}{(2 \cdot 1)(2 \cdot 1)} = \frac{720}{4} = 180$$

The number of permutations, P, of n elements, taken n at a time, with n_1 elements alike, and n_2 of another kind alike is:

$$\frac{n!}{n_1!\ n_2!}$$

Exercises

Find the number of permutations of all the letters in each word.

1. Deeded
2. Cincinnati
3. Weeded
4. How many different permutations can be made by flying six pennants at a time on a vertical flag pole, if two are red, three are blue, and one is white?

Answers to Exercises

1. $\frac{6!}{3!3!} = \frac{6 \cdot 5 \cdot 4 \cdot \not{3} \cdot \not{2} \cdot \not{1}}{3 \cdot 2 \cdot 1 \cdot \not{3} \cdot \not{2} \cdot \not{1}} = \frac{6 \cdot 5 \cdot 4}{3 \cdot 2 \cdot 1} = \frac{120}{6} = 20$

2. $\frac{10!}{2!3!3!} = \frac{10 \cdot 9 \cdot 8 \cdot 7 \cdot 6 \cdot 5 \cdot 4 \cdot \not{3} \cdot \not{2} \cdot \not{1}}{2 \cdot 1 \cdot 3 \cdot 2 \cdot 1 \cdot \not{3} \cdot \not{2} \cdot \not{1}} = \frac{604{,}800}{12} = 50{,}400$

3. $\frac{6!}{2!3!} = \frac{6 \cdot 5 \cdot 4 \cdot \not{3} \cdot \not{2} \cdot \not{1}}{2 \cdot 1 \cdot \not{3} \cdot \not{2} \cdot \not{1}} = \frac{120}{2} = 60$

4. $\frac{6!}{2!3!} = \frac{6 \cdot 5 \cdot 4}{2} = 60$

COMBINATIONS

Arrangements without consideration of the order of elements are called *combinations*. There are six permutations of three letters each using the letters a, b, and c:

$$P = (a, b, c)\ (a, c, b)\ (b, a, c)\ (b, c, a)\ (c, b, a)\ (c, a, b)$$

But there is just one combination (a, b, c).

The combination of two elements from one set $[a, b, c]$ are:

$$(a, b)$$
$$(b, c)$$
$$(a, c)$$

The symbol C indicates this is a combination. The combinations can be represented as:

$${}_3C_2$$

That is, the combination of two elements from a set of three elements.

From each combination, two permutations can be formed, or 2^P2.

Combinations	**Permutations**
(a, b)	$(a, b)\ (b, a)$
(b, c)	$(b, c)\ (c, b)$
(a, c)	$(a, c)\ (c, a)$

Example

A club of twelve members wishes to choose a president, vice president, and secretary. The order in which the members' names are written on a ballot will determine who will fill each office. How many different selections are possible?

The question deals with the number of ways each element of one set:

SET (Club Members)
1, 2, 3, 4, 5, 6, 7, 8, 9, 10, 11, 12

can be matched with each element of the subset:

SUBSET (Officers)
1 (president), 2, (vice president), 3, (secretary)

There are twelve choices for president. But once a member has been listed on a ballot as president, he cannot appear on the same ballot as vice president. The order of selection is important. Once a member has been written in as president, there are only eleven possibilities left for vice president. Once members have been chosen as president and vice president, there are only ten possibilities left for secretary.

The number of possible selections (permutations) are:

$$_{12}P_3 = 12 \cdot 11 \cdot 10 = 1{,}320$$

In a second situation, three committee members are to be selected from a club of twelve members. The order of selection is *not important*. The selection is called a combination. The twelve indicates the number of people and the three indicates the number of people to be selected.

A combination of n elements taken r at a time is written as nCr.

nCr is defined as $\dfrac{n}{r!(n-r)!}$

That is:

1. The factorial of the total elements
2. Divided by the factorial of the elements taken at one time
3. Multiplied by the factorial of the number left over

A combination of twelve elements taken three at a time is:

$$_{12}C_3 = \frac{12!}{3!(12-3)!} = \frac{12 \cdot 11 \cdot 10 \cdot \cancel{9} \cdot \cancel{8} \cdot \cancel{7} \cdot \cancel{6} \cdot \cancel{5} \cdot \cancel{4} \cdot \cancel{3} \cdot \cancel{2} \cdot \cancel{1}}{3 \cdot 2 \cdot 1 \cdot \cancel{9} \cdot \cancel{8} \cdot \cancel{7} \cdot \cancel{6} \cdot \cancel{5} \cdot \cancel{4} \cdot \cancel{3} \cdot \cancel{2} \cdot \cancel{1}}$$

The like numbers in the denominator cancel like numbers in the numerator and we are left with:

$$_{12}C_3 = \frac{12 \cdot 11 \cdot 10}{3 \cdot 2 \cdot 1} = 220$$

Exercises with Combinations

Solve the following:

1. $_7C_4 =$

2. $_4C_3 =$

3. $_{12}C_2 =$

4. $_{12}C_7 =$

5. How many ways can three prizes be awarded to a group of eight people?

Answers

1. $_7C_4 = \dfrac{7!}{4!3!} = \dfrac{7 \cdot 6 \cdot 5 \cdot 4 \cdot \cancel{3} \cdot \cancel{2} \cdot \cancel{1}}{4 \cdot 3 \cdot 2 \cdot 1 \cdot \cancel{3} \cdot \cancel{2} \cdot \cancel{1}} = \dfrac{840}{24} = 35$

2. $_4C_3 = \dfrac{4!}{4!1!} = \dfrac{4 \cdot 3 \cdot 2 \cdot \cancel{1}}{4 \cdot 3 \cdot 2 \cdot 1 \cdot \cancel{1}} = \dfrac{24}{24} = 1$

3. $_{12}C_2 = \dfrac{12!}{2!10!} = \dfrac{12 \cdot 11 \cdot \cancel{10} \cdot \cancel{9} \cdot \cancel{8} \cdot \cancel{7} \cdot \cancel{6} \cdot \cancel{5} \cdot \cancel{4} \cdot \cancel{3} \cdot \cancel{2} \cdot \cancel{1}}{2 \cdot 1 \cdot \cancel{10} \cdot \cancel{9} \cdot \cancel{8} \cdot \cancel{7} \cdot \cancel{6} \cdot \cancel{5} \cdot \cancel{4} \cdot \cancel{3} \cdot \cancel{2} \cdot \cancel{1}} = \dfrac{132}{2} = 66$

4. $_{12}C_7 = \dfrac{12!}{7!5!} = \dfrac{12 \cdot 11 \cdot 10 \cdot 9 \cdot 8 \cdot \cancel{7} \cdot \cancel{6} \cdot \cancel{5} \cdot \cancel{4} \cdot \cancel{3} \cdot \cancel{2} \cdot \cancel{1}}{\cancel{7} \cdot \cancel{6} \cdot \cancel{5} \cdot \cancel{4} \cdot \cancel{3} \cdot \cancel{2} \cdot \cancel{1} \cdot 5 \cdot 4 \cdot 3 \cdot 2 \cdot 1} = \dfrac{95040}{120} = 792$

5. $_8C_3 = \dfrac{8!}{3!5!} = \dfrac{8 \cdot 7 \cdot 6 \cdot \cancel{5} \cdot \cancel{4} \cdot \cancel{3} \cdot \cancel{2} \cdot \cancel{1}}{3 \cdot 2 \cdot 1 \cdot \cancel{5} \cdot \cancel{4} \cdot \cancel{3} \cdot \cancel{2} \cdot \cancel{1}} = \dfrac{336}{6} = 56$

Practice Problems: Probabilities, Combinations, and Permutations

1. In a certain registration district, 18 boys and 21 girls were born during one month. Find the probability that the first child born during that month was a boy.

2. The probability that Washington High will win tonight's basketball game is $2/5$ and the probability that Kennedy High will win their game tonight is $3/7$. Find the probability that both Washington and Kennedy will win tonight.

3. A coin is tossed ten times. Find the most probable number of times heads will turn up.

4. John and Jose are running in different races. The probability that John will win his race is $1/2$. The probability that Jose will win his race is $1/5$. What is the probability that both will win?

5. If ten runners compete in a race, in how many different ways can prizes be awarded for first, second, and third places?

6. In how many different orders can you arrange five videos on a shelf?

7. In how many ways can four people be seated in twelve chairs?

8. In a school election, there are two candidates for president, three for vice president, four for secretary, and one for treasurer. How many ways may the election result?

9. On three shelves in the library, there are 15 books, 14 books, and 20 books, respectively. In how many ways may you choose one book from each shelf?

10. A basketball coach must choose a center, left guard, right guard, left forward, and a right forward from a team of ten players. How many ways can this be done?

11. How many pairs of letters are possible if each pair contains one of the vowels a, i, u, and one of the consonants b, c, d, f, and g?

12. A man can choose one of four envelopes. One envelope contains $100 and the others are empty. What is the probability that he will choose the $100 envelope?

13. How many distinguishable permutations can be made from the word Mississippi?

14. How many positive odd integers less than 1,000 can be formed using the digits 2, 3, 4, 5, 6, and 8?

15. One marble is drawn from a bag of three black, four white, and six red marbles. Find the probability that the marble is black.

16. Find the number of distinguishable ways that three identical TV sets and identical record players can be arranged side by side.

17. How many different permutations can be made by flying six pennants at a time on a vertical flag pole if they are all different colors?

18. How many different permutations can be made by flying six pennants at a time on a vertical flag pole if two are red, three are blue, and one is white?

19. Find the number of committees of four persons that can be selected from a set of 15 people?

20. How many ways can five of eight potential jurists be seated in five vacant seats in court?

Answers to Practice Problems: Probabilities, Permutations, and Combinations

1.

$$18 \text{ boys} + 21 \text{ girls} = 39 \text{ born}$$

$$\frac{18 \text{ boys}}{39 \text{ total}} = \frac{6}{13}$$

2.

$$2/5 \cdot 3/7 = 6/35$$

3.

$$\frac{10 \textit{ tosses}}{2 \textit{ possibilities}} = 5$$

4.

$1/2 \cdot 1/5 = 1/10$ probability both will win their races

5.

$10 \cdot 9 \cdot 8 = 720$ different ways

6.

$5 \cdot 4 \cdot 3 \cdot 2 \cdot 1 = 120$ ways

7.

$12 \cdot 11 \cdot 10 \cdot 9 = 11{,}880$ ways in which four people can be seated in twelve chairs

8.

$$2 \times 3 \times 4 \times 1 = 24 \text{ ways}$$

9.

$$15 \times 14 \times 20 = 4{,}200$$

10.

$$10 \cdot 9 \cdot 8 \cdot 7 \cdot 6 = 30{,}240 \text{ ways}$$

11.

$$3 \text{ vowels} \times 5 \text{ consonants} = 15 \text{ pairs}$$

12.

$$P = \tfrac{1}{4}$$

13.

$$\frac{11!}{4!4!2!} = \frac{11 \cdot 10 \cdot 9 \cdot 8 \cdot 7 \cdot 6 \cdot 5 \cdot \cancel{4} \cdot \cancel{3} \cdot \cancel{2} \cdot \cancel{1}}{4 \cdot 3 \cdot 2 \cdot 1 \cdot \cancel{4} \cdot \cancel{3} \cdot \cancel{2} \cdot \cancel{1}} = \frac{1663200}{48} = 34{,}650$$

14.

$$\begin{aligned} \underline{2} &= 2 \\ \underline{6} \cdot \underline{2} &= 12 \\ \underline{6} \cdot \underline{6} \cdot \underline{2} &= \underline{72} \\ & 86 \end{aligned}$$

15.

$$P(a) = \frac{n(a)}{n(s)} = \frac{3}{13}$$

16.

$$\frac{6!}{3!3!} = 20$$

17.

$$6! = 720$$

18.

$$\frac{6!}{2!3!} = \frac{6 \cdot 5 \cdot 4 \cdot \cancel{3} \cdot \cancel{2} \cdot \cancel{1}}{2 \cdot 1 \cdot \cancel{3} \cdot \cancel{2} \cdot \cancel{1}} = \frac{120}{2} = 60$$

19.

$${}_{15}C_4 = \frac{15!}{4!11!} \; \frac{15 \cdot 14 \cdot 13 \cdot 12 \cdot \cancel{11} \cdot \cancel{10} \cdot \cancel{9} \cdot \cancel{8} \cdot \cancel{7} \cdot \cancel{6} \cdot \cancel{5} \cdot \cancel{4} \cdot \cancel{3} \cdot \cancel{2} \cdot \cancel{1}}{4 \cdot 3 \cdot 2 \cdot 1 \cdot \cancel{11} \cdot \cancel{10} \cdot \cancel{9} \cdot \cancel{8} \cdot \cancel{7} \cdot \cancel{6} \cdot \cancel{5} \cdot \cancel{4} \cdot \cancel{3} \cdot \cancel{2} \cdot \cancel{1}} = \frac{32760}{24} = 1365$$

20.

$${}_8P_5 = 8 \cdot 7 \cdot 6 \cdot 5 \cdot 4 = 6{,}720$$

TABLE 1

TABLE OF SQUARES AND SQUARE ROOTS

N	N^2	$\sqrt{N}$	N	N^2	$\sqrt{N}$
1	1	1	51	2,601	7.141
2	4	1.414	52	2,704	7,211
3	9	1.732	53	3,809	7.280
4	16	2	54	2,916	7.348
5	25	2.236	55	3,025	7.416
6	36	2.449	56	3.136	7.483
7	49	2.646	57	3,249	7.550
8	64	2.828	58	3.364	7.616
9	81	3	59	3,481	7,681
10	100	3.162	60	3,600	7.746
11	121	3.317	61	3,721	7.810
12	144	3.464	62	3,844	7.874
13	169	3.606	63	3,969	7.937
14	196	3.742	64	4,096	8
15	225	3.873	65	4,225	8.062
16	256	4	66	4,356	8.124
17	289	4.123	67	4,489	8.185
18	324	4.243	68	4,624	8.246
19	361	4.359	69	4,761	8.307
20	400	4.472	70	4,900	8.367
21	441	4.583	71	5,041	8.426
22	484	4.690	72	5,184	8.485
23	529	4.796	73	5,329	8.544
24	576	4.899	74	5,476	8.602
25	625	5	75	5,625	8.660
26	676	5.099	76	5,776	8.718
27	729	5.196	77	5,929	8.775
28	784	5.292	78	6,084	8.832
29	841	5.385	79	6,241	8.888
30	900	5.477	80	6,400	8.944
31	961	5.568	81	6,561	9
32	1,024	5.657	82	6,724	9.055
33	1,089	5.745	83	6,889	9.110
34	1,156	5.831	84	7,056	9.165
35	1,225	5.916	85	7,225	9.220
36	1,296	6	86	7,396	9.274
37	1,369	6,083	87	7,569	9.327
38	1,444	6.164	88	7,744	9.381
39	1,521	6.245	89	7,921	9.434
40	1,600	6.325	90	8,100	9.487
41	1,681	6.403	91	8,281	9.539
42	1,764	6.481	92	8,464	9.592
43	1,849	6.557	93	8,649	9.644
44	1,936	6.633	94	8,836	9.695
45	2,025	6.708	95	9,025	9.747
46	2,116	6.782	96	9,216	9.798
47	2,209	6.856	97	9,409	9.849
48	2,304	6.928	98	9,604	9.899
49	2,401	7	99	9,801	9.950
50	2,500	7.071	100	10,000	10

TABLE 2

TABLE OF VALUES OF TRIGONOMETRIC FUNCTIONS

deg	sin	cos	tan	deg	sin	cos	tan	deg	sin	cos	tan
0	.000	1.000	.000								
1	.017	1.000	.017	31	.515	.857	.601	61	.875	.485	1.804
2	.035	.999	.035	32	.530	.848	.625	62	.883	.470	1.881
3	.052	.999	.052	33	.545	.839	.649	63	.891	.454	1.963
4	.070	.998	.070	34	.559	.829	.675	64	.899	.438	2.050
5	.087	.996	.087	35	.574	.819	.700	65	.906	.423	2.145
6	.105	.995	.105	36	.588	.809	.727	66	.914	.407	2.246
7	.122	.993	.123	37	.602	.799	.754	67	.921	.391	2.356
8	.139	.990	.141	38	.616	.788	.781	68	.927	.375	2.475
9	.156	.988	.158	39	.629	.777	.810	69	.934	.358	2.605
10	.174	.985	.176	40	.643	.766	.839	70	.940	.342	2.747
11	.191	.982	.194	41	.656	.755	.869	71	.946	.326	2.904
12	.208	.978	.213	42	.669	.743	.900	72	.951	.309	3.078
13	.225	.974	.231	43	.682	.731	.933	73	.956	.292	3.271
14	.242	.970	.249	44	.695	.719	.966	74	.961	.276	3.487
15	.259	.966	.268	45	.707	.707	1.000	75	.966	.259	3.732
16	.276	.961	.287	46	.719	.695	1.036	76	.970	.242	4.011
17	.292	.956	.306	47	.731	.682	1.072	77	.974	.225	4.331
18	.309	.951	.325	48	.743	.669	1.111	78	.978	.208	4.705
19	.326	.946	.344	49	.755	.656	1.150	79	.982	.191	5.145
20	.342	.940	.364	50	.766	.643	1.192	80	.985	.174	5.671
21	.358	.934	.384	51	.777	.629	1.235	81	.988	.156	6.314
22	.375	.927	.404	52	.788	.616	1.280	82	.990	.139	7.115
23	.391	.921	.424	53	.799	.602	1.327	83	.993	.122	8.144
24	.407	.914	.445	54	.809	.588	1.376	84	.995	.105	9.514
25	.423	.906	.466	55	.819	.574	1.428	85	.996	.087	11.430
26	.438	.899	.488	56	.829	.559	1.483	86	.998	.070	14.301
27	.454	.891	.510	57	.839	.545	1.540	87	.999	.052	19.081
28	.470	.883	.532	58	.848	.530	1.600	88	.999	.035	28.636
29	.485	.875	.554	59	.857	.515	1.664	89	1.000	.017	57.290
30	.500	.866	.577	60	.866	.500	1.732	90	1.000	.000	-